Signals and Communication Technology

More information about this series at http://www.springer.com/series/4748

Hermann Merz · Thomas Hansemann
Christof Hübner

Building Automation

Communication Systems with EIB/KNX, LON and BACnet

Second Edition

Springer

Hermann Merz
Fakultät für Elektrotechnik
Hochschule Mannheim
Mannheim
Germany

Christof Hübner
Fakultät für Elektrotechnik
Hochschule Mannheim
Mannheim
Germany

Thomas Hansemann
Fakultät für Elektrotechnik
Hochschule Mannheim
Mannheim
Germany

Translated by: James Backer, Heidelberg, Germany; Viktoriya Moser, Rüsselsheim, Germany; Leena Greefe, Schönau, Germany; Benjamin Neus, Mannheim, Germany

ISSN 1860-4862 ISSN 1860-4870 (electronic)
Signals and Communication Technology
ISBN 978-3-030-10336-1 ISBN 978-3-319-73223-7 (eBook)
https://doi.org/10.1007/978-3-319-73223-7

Original German edition published by Hanser Fachbuchverlag, 2007

Softcover re-print of the Hardcover 2nd edition 2018

Printed on acid-free paper

This Springer imprint is published by the registered company Springer International Publishing AG part of Springer Nature
The registered company address is: Gewerbestrasse 11, 6330 Cham, Switzerland

Foreword 2nd Edition

An ever-increasing number of processes are being automated in our modern industrial society. The level of automation in residential and commercial buildings is also steadily increasing due to the demand for more comfort, convenience, security and efficiency. Building automation has become an integral part of automation engineering and offers customized solutions for all types of buildings. In order to process their sometimes very complex functions, the employed sensors, actuators, controllers/regulators and visualizations require suitable industrial communication systems to exchange data among themselves or with external systems. Therefor field buses and computer networks are employed. This book will provide you with an introduction to building automation and building systems engineering and a detailed insight into the following topics:

- Usage of DDC automation devices and energy management functions
- The basics of industrial communication technology
- European Installation Bus (KNX)
- Local Operating Network (LON)
- TCP/IP computer networks and the communication protocol BACnet

We would like to thank all readers for their constructive feedback on the 1. edition of this book. That way, we had the chance to learn that many schools, trade schools, universities of applied science and other universities use our book. We appreciate that a lot. In the present 2. edition many illustrations were updated, a subchapter on usage and planning dimensioning of DDC automation devices was added to chapter 1 and chapter 3 was adjusted to the commissioning software ETS 5.

Solutions to the exercises in this book are available.

Mannheim, Germany
March 2018

Thomas Hansemann
Christof Hübner
Hermann Merz

Contents

Chapter 1
Introduction to Building Automation

1.1 What Is Building Automation?

The level of automation in residential and commercial buildings has risen steadily over the years. This is not only due to the increasing demand for more comfort and convenience, but also the benefits building automation brings with regard to saving and managing energy. Security is another important factor, particularly in residential buildings. Whereas in commercial buildings flexibility is high on the agenda —offices buildings, for example, should be designed in such way that they can be easily adapted to meet any change in use or requirements.

1.1.1 Building Automation in Private Residential Buildings

A variety of automated functions are commonplace in many modern residential buildings. One of the most obvious examples is the use of control functions in heating systems for the optimal regulation of energy consumption. Today all new installations have sophisticated combustion controllers and room temperature regulators (thermostats). These thermostats usually come with an in-built timer-switch program for automatically reducing the room temperature at night. These programs have become standard features because they are compatible with a large number of applications, and therefore operate from the word go without the need for any additional programming or configuration.

Automatic lighting control is another example of automation in residential buildings. Exterior lights are often connected to motion detectors, so that they come on automatically should someone approach. Motion detectors detect the heat radiation of an approaching person. Combined with a brightness sensor, this ensures that the light only comes on if it is dark enough. Even though this is a comparatively simple automation function, it nevertheless illustrates the

H. Merz et al., *Building Automation*, Signals and Communication Technology,
https://doi.org/10.1007/978-3-319-73223-7_1

combination of event control and logical connections. This example focuses on comfort and ease of use.

A more complex example of automation involves being able to turn all the lights in a house on or off from one central point-particularly useful if you hear an intruder at night. To achieve this with a conventional electrical installation requires an immense amount of wiring, because each lamp needs to have its own wire connecting it directly to the one switch. By connecting all the light switch components to a bus system over which they can communicate, you do not need as much wiring, making it easier and more affordable to implement this panic button function. The focus here is on security.

In summary, automation in private residential buildings focuses on:

- Cost effectiveness/saving energy
- Comfort and convenience
- Security

1.1.2 Building Automation in Commercial Buildings

Commercial buildings within the context of building automation are buildings that serve a purely functional purpose, for example, offices, shopping centers, hospitals, railway stations, airport terminals and underground car parks.

In modern buildings there are a variety of automation systems for heating, ventilating and air conditioning (see Fig. 1.1). To ensure these systems run smoothly and economically, they are fitted with sophisticated controllers, which are often interconnected with each other and to a control center via field buses and networks. These control systems optimize energy consumption and enable support and maintenance personnel to carry out their jobs more efficiently.

Studies carried out on the workplace have shown that staff performance and productivity is at its highest in a comfortable environment and drops considerably if, for example, the temperature in an office is too high during the summer. This has led to the installation of air-conditioning systems in an increasing number of offices in new commercial buildings. Even the way we operate these systems has changed. Today blinds and lights can be controlled from office computers, increasing comfort and usability and, in so doing, optimizing employee productivity and performance [10].

Systems in commercial buildings must be flexible. If a company wants to restructure the layout of an office by converting a large conference room into a number of smaller offices, the layout and set up of the building's operational equipment must enable these changes. Building automation systems enable you to connect a light switch to a light by simply reprogramming the intelligent components, rather than rewiring the electrics. The focus here is on flexibility.

Fig. 1.1 A ventilation system in a commercial building [1]

In summary, automation in commercial buildings focuses on:

- Cost-effectiveness/saving energy
- Communication via bus systems and networks
- Comfort and convenience
- Flexibility.

1.2 The Difference Between Building Automation and Building Control

When we talk about automated functions in buildings, the terms "building automation" and "building control" are often used. At first glance these terms appear synonymous. For clarification the Association of German Engineers (*Verein Deutscher Ingenieure*) defines building automation as follows:

Building automation is the computerized measurement, control and management of building services [11].

From this definition we can deduce that building control is a part of building automation. Building automation was first implemented in commercial buildings to enable functions to run automatically. This also included the first use of direct digital controllers (DDCs) (Fig. 1.2) in heating, ventilation and air-conditioning systems (HVACs). Furthermore, by using a control center you can operate and monitor the systems more effectively, and also create a cross-system network.

Building control is a specific subdivision of building automation that focuses mainly on electrical installations.

Building control refers to the use of an installation bus to connect system components and devices to a system designed for a specific electrical installation that controls and connects all the functions and processes in a building. All of the components have their own "intelligence" and exchange information directly with each other [4].

Building control components, such as four-gang blind actuators (see Fig. 1.3), are usually mounted in a control cabinet or next to the device to be controlled (for example, a blind). Building control systems do not require central DDCs to process control or regulation functions.

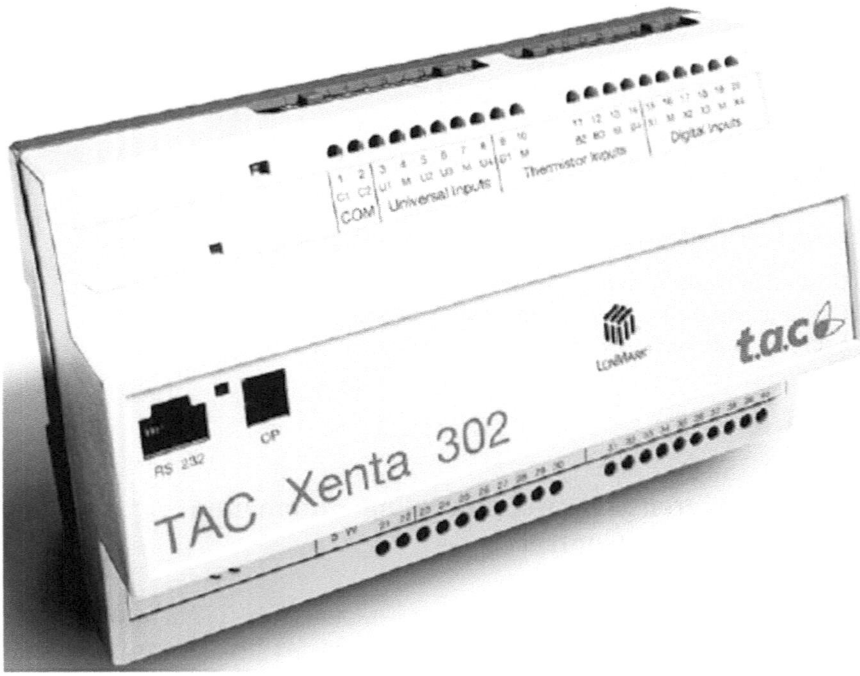

Fig. 1.2 A direct digital controller (DDC) [5]

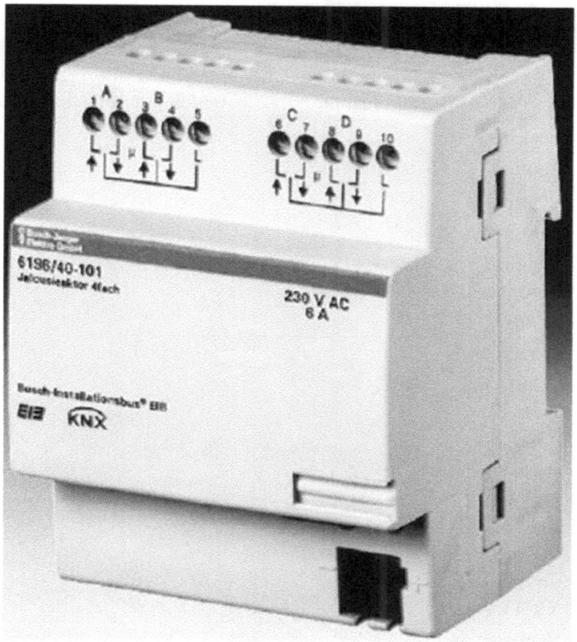

Fig. 1.3 A 4-gang blind actuator for mounting in a control cabinet [Busch-Jaeger Elektro]

1.2.1 Systems in Building Automation

Building service equipment includes all the systems necessary for a building to operate, the most important of which are the heating, ventilation, air conditioning, water and electrical energy supply, and sanitation systems such as sewage pumping stations.

DDCs are now essential because the operational processes in a building must operate automatically if they are to be cost effective. The vendor who supplies the DDCs for specific systems is responsible for measuring and controlling (MC) these systems-primarily HVAC. Table 1.1 summarizes the systems in building automation.

Building automation involves coordinating and connecting all the systems in a building, so that they can communicate with each other. This can be achieved in three ways:

- Systems can be connected via DDCs and building control components. This is common in heating, ventilation, air-conditioning, lighting and shade control systems.
- Systems can also be connected via special DDCs that perform only input and output functions. This is common in sanitation and power supply systems that have their own in-built automation mechanisms.

Table 1.1 Systems in building automation

System	Usually integrated into building automation	Increasingly integrated into building automation	Systems that are controlled by DDCs or other building automation components
Heating	×		×
Cooling	×		×
Ventilation	×		×
Power supply	×		
Lighting control	×		×
Blinds	×		×
Sanitation	×		
Central fire alarm	×		
Burglar alarm		×	
Access control		×	
Video surveillance (CCTV)		×	
Network engineering		×	
Multimedia		×	
Elevators		×	
Telephones		×	
Maintenance management		×	
Payroll/ accounting		×	
Facility management		×	

- If a system needs to transfer a large amount of information or has its own computer, then it can be directly connected to the building automation control computer. Data is then transferred via a bus system or network as opposed to over individual wires. This is common in subordinate video or superordinate accounting systems.

The interfaces between the individual operational systems of each facility must always be clearly defined in terms of data exchange and logistics.

In building automation, information technology is used to link all the systems in a building, enabling them to be centrally monitored by a control computer at the management level (see Fig. 1.4). In older systems, the control computer also processes regulation functions. This is called building management system BMS.

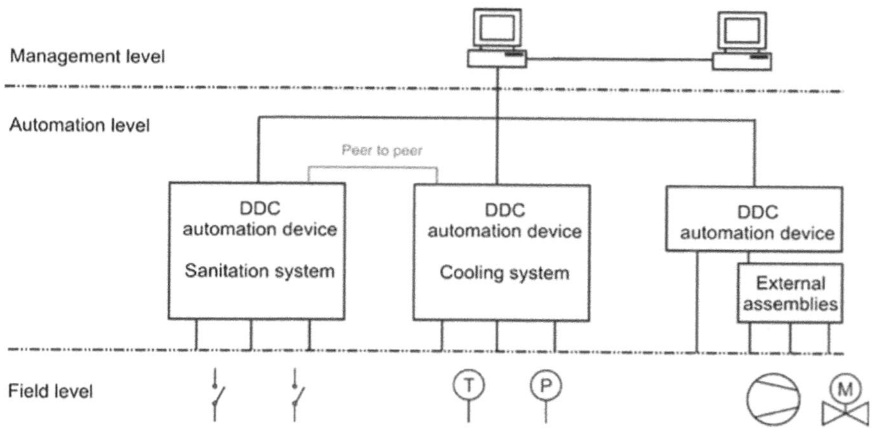

Fig. 1.4 The IT network of systems in building automation

Information exchange between the individual systems generally occurs at the automation level. Information is transferred over so-called peer-to-peer connections, which are logical pathways that use physical bus or network connections.

Another feature is the external assembly seen at the bottom right. In this case the regulation functions remain in the DDC automation device but the connected assembly processes certain parts of the function individually. Examples for this are frequency inverters for large fans used to control the rate of rotation or electronic meters used to measure energy consumption. In any case the external assembly is connected to the DDC automation device by bus.

1.2.2 *Systems in Building Control*

Building control represents a small subsection of building automation and involves the localized automation of components in an individual room—known as single room control or room automation (see Table 1.2). The building control components combine all the functions that are needed to provide a comfortable and energy-saving environment. DDCs are not used because the functions are distributed across the intelligent building control components.

Table 1.2 Systems in building control

System	Room automation possible with building control components
Heating, cooling, and ventilation	×
Lighting control	×
Shade/blinds	×

Fig. 1.5 A building control switch actuator [3]

The individual components for each application are pre-programmed for specific tasks. For example, an intelligent processor-controlled push button directly connected to the bus is used to send the signal to turn on a light. Another component is then used as an intelligent processor-controlled switch actuator to execute the command (see Fig. 1.5). This actuator is either mounted directly next to the light or in a control cabinet.

These types of components are also used to control and regulate radiators. An electronic actuator is installed in the radiator and is connected, via the bus, to a temperature sensor near the door. The beauty of this solution is that it enables you to easily connect the various systems in the room. For example, by installing a presence sensor near the door, you can ensure that as the last person leaves the room, the lights are automatically switched off and the radiator is turned down or off. The automated functions are processed by the building control components and not by a central DDC.

Figure 1.6 gives you an idea of the building control systems found in a room.

1.3 The Structure of Building Automation and Control Networks

1.3.1 The Hierarchical Structure of Building Automation

The components required for processing control functions in automated systems are organized hierarchically. Figure 1.7 shows the typical architecture found in building automation.

Right next to the process are the sensors necessary for recording system information in building automation. These include temperature sensors, flow meters and status registration devices (e.g. frost detectors). You will also find actuators for controlling output commands to the operational system interface.

Ventilation systems have valves that regulate the flow rate of the heating circuit, or drives that control the flaps to regulate the amount of air drawn in from outside. Figure 1.8 shows the sensors and actuators (so-called assemblies) in a ventilation system.

Wires connect the sensors and actuators to the DDCs that control and regulate the system(s). For each information such as status messages or sensor signals one

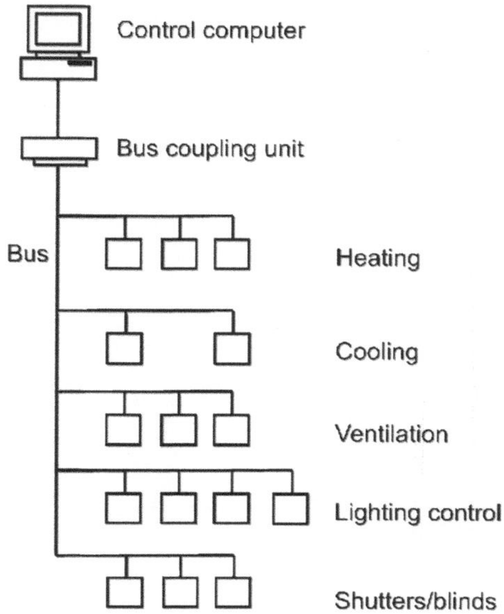

Fig. 1.6 Building control systems in a room

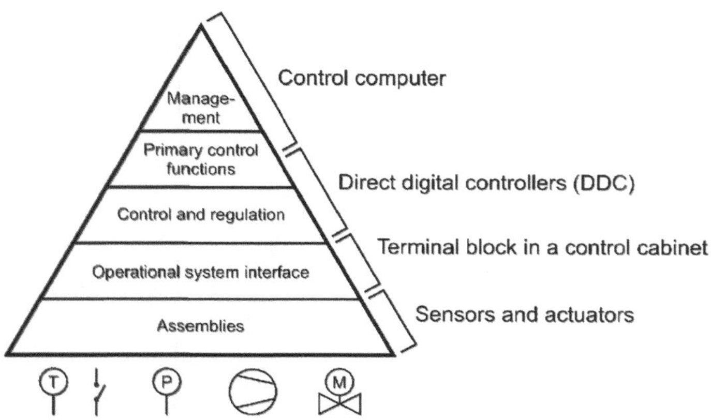

Fig. 1.7 The hierarchical structure in building automation

pair of wires is used. The DDCs are mounted in a control cabinet (see Fig. 1.9), which is positioned next to the technical system. The close proximity of the control cabinet to the technical system reduces the amount of cabling required. Even a standard ventilation system installed in a commercial building needs ~1.2 km of cable to send and receive 40 information messages. A terminal block is housed in the control cabinet and is used to connect the cables to the technical system, which is why it is called the system interface.

Fig. 1.8 Sensors and actuators in a ventilation system [1]

Fig. 1.9 The terminal block and DDCs mounted in a control cabinet [1]

The DDCs housed in the control cabinet process all control functions and, therefore, enable the whole system to operate automatically. The DDCs do not need to be connected to a central control computer. Even at the automation level the

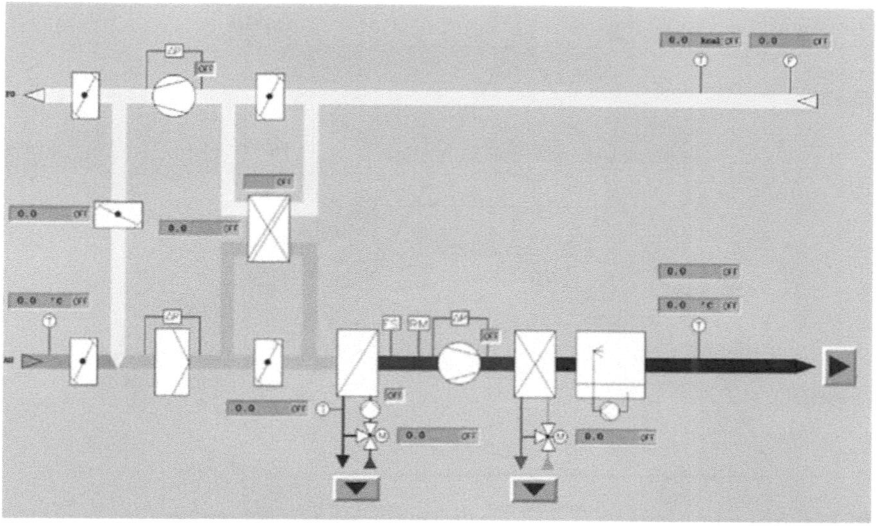

Fig. 1.10 A ventilation system displayed on a control computer

DDCs have energy-saving programs integrated into their software, for example, for controlling the position of the ventilation flaps to let in the desired amount air from the outside based on the temperature in the room.

If all the systems are in close proximity to each other and the building operator does not have to make constant adjustments, then specially optimized DDC can be used to implement high-level control functions. Alternatively, these high-level control functions can be managed by a control computer (see Fig. 1.10).

Control computers can also execute cross-system functions, because the systems transfer all information over the same transmission medium. An excellent example of this is a timer-switch program set to the times the building is in use. This program automatically shuts down all non-essential systems in the evening and starts them up again in the morning.

The control computer also runs all the necessary building management programs, including logging all events and alarms, archiving all measured readings, and graphically displaying the status of the operational systems. It also forwards information to other computer systems, for example, energy and load meter readings to a main accounting system.

1.3.2 The Hierarchical Structure in Building Control

By housing the sensor together with an in-built processor and a bus connector, you can combine different levels into one (see Fig. 1.11).

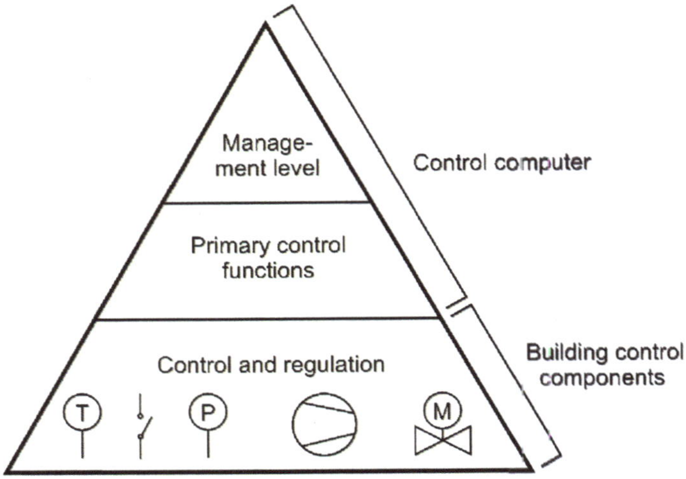

Fig. 1.11 The special hierarchical structure found in building control systems

Fig. 1.12 A building control temperature sensor with a setpoint adjuster and control program (a Busch-triton® 5-gang switch sensor with thermostat) [Busch-Jaeger Elektro]

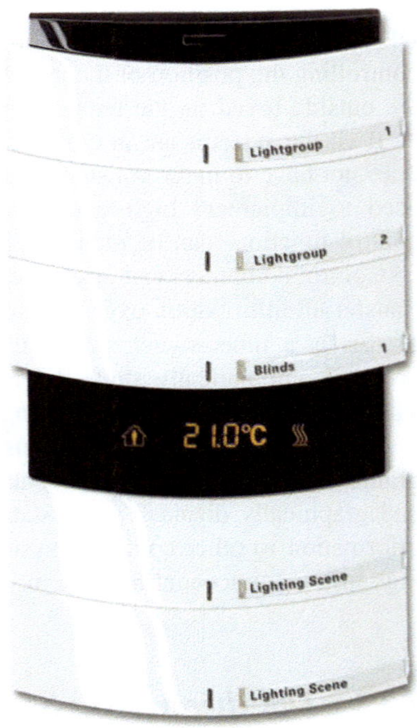

Figure 1.12 shows a device comprising a five-gang switch sensor and a room temperature controller (Busch-triton®). This device contains an in-built sensor and processor. The sensor sends its temperature reading to the processor, which then processes the information. It also allows you to set and control the (programmable) setpoint for the room temperature.

The five-gang switch sensor can send, for example, switching, dimming, blind, measurement, or ventilation messages. The top three rocker switches are assigned to the room temperature controller. The bottom two rocker switches can be used to control the lighting scenes. The integrated LCD displays the actual room temperature, the setpoint temperature and the operating mode.

The operational system interface, shown in Fig. 1.7, is not visible from the outside. Furthermore, the in-built microcontroller controls the system directly. This device regulates the room temperature by comparing the setpoint temperature with the room temperature reading, and then sends the controller's output signal via a bus to the heater's in-built electrical actuator.

1.4 Usage of DDC Automation Devices

This subchapter explains the input and output functions of DDC automation devices as part of building automation. Based on a ventilation system the assignment of the device to an operational system is shown in an exemplary way. The chapter is concluded with an overview of the usual scope of services and supplies for a turnkey system.

1.4.1 Basic Functions of Building Automation

The purpose of the DDC automation devices is to monitor, control and regulate the previously named systems (see 1.2.1) independently. To do so, the information from the systems is connected to the DDC by using sensors and actuators. This connection can be either a physical connection using a wire or a communicative connection using a bus. The amount of necessary input and output functions of the DDC highly depends on the size of the system that is to be connected. However, since the different sensors and actuators are very diverse, an adjustment of the automation device to different forms of signals should be considered as well.

The assignment of these functions and their quantity to operational systems is accomplished by using an information list as seen in Fig. 1.13 in accordance with VDI 3814 Part 1 and EN ISO 16484 [12]. The colloquial name of an information list of this kind is data point list. Columns 1–5 of section 1 contain information on which physical input and output functions the required DDCs need to have. The functions mentioned here are called basic functions. The functions mentioned in section 2 refer to a connection of sensors and actuators by bus instead of an

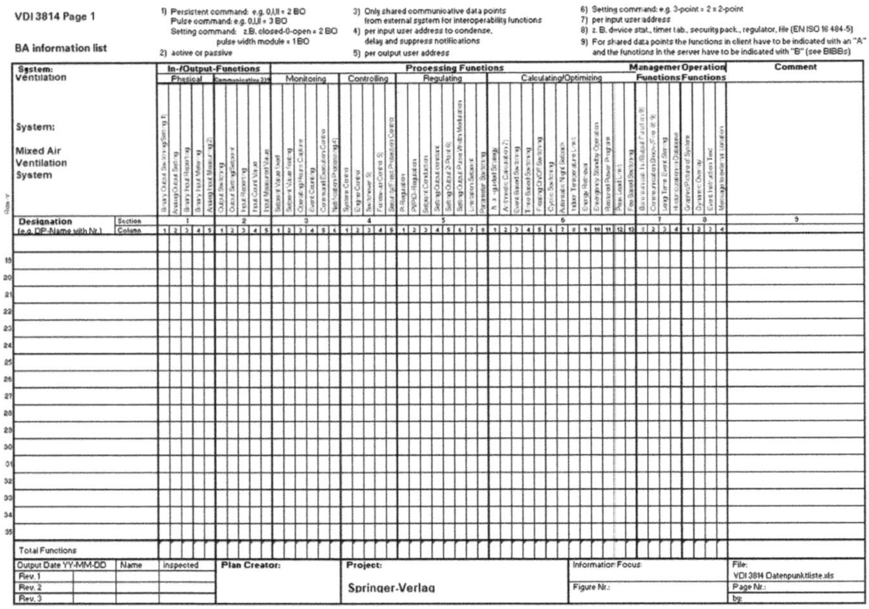

Fig. 1.13 Information list in accordance with VDI 3814 part 1 and EN ISO 16484

individual connection of every aggregate by wire. For cost efficiency reasons, the individual signals are usually connected by wire.

The additional columns in the information list contain information on the planned processing functions of the DDC for monitoring, controlling, regulating and optimizing (sections 3–6). Many of the energy management functions described in section 5 can be found in section 6. Finally, the last columns are for management and operation functions of the control computer (sections 7–8) [2].

1.4.1.1 Basic Function Reporting

The purpose of the basic function reporting is the status detection of information from the operational system. The respective inputs of the DDC automation device are called binary input (BI) or digital input (DI).

Essentially, two kinds of reports can be distinguished for the function reporting. Status and maintenance notifications are gathered with a normally open contact as seen in Fig. 1.14. Since this contact is open by default, the circuit is called normally open (n. o.).

Fig. 1.14 Basic function reporting by use of normally open contact

Fig. 1.15 Basic function reporting by use of normally closed contact for wire break control

Particularly vital information like malfunctioning or alarm notifications are gathered using a normally closed contact as seen in Fig. 1.15. Since this contact is closed by default, the circuit is called normally closed (n. c.). The benefit of this layout is an additional wire break control that can lead to an alarm as well.

1.4.1.2 Basic Function Metering

The purpose of the basic function metering is to gather count values such as values of electricity meters, water meters or heat meters. The respective inputs of the DDC automation device are special binary inputs that are equipped with a conversion function within the device. Impulses provided by meters over potential-free contacts are captured here and converted into energy values. Generic electrical meters provide e.g. 9 impulses per kWh that can be processed for energy data acquisition or for accounting purposes.

The disadvantage of gathering count values by means of a physical wire connection is that a disruption through wire break leads to a difference in the values displayed by the meter and the ones captured by the DDC. In this case, an equalization within the DDC program is necessary. Therefore, the basic function metering is usually realized with a communicative connection of an electrical meter with bus connection (cf. Fig. 1.4 bottom right). Should a disruption of signal occur now, the system can automatically update itself afterwards.

1.4.1.3 Basic Function Measuring

The purpose of the basic function measuring is the capture of steady analog measurement signals from the operational system. The respective input of the DDC automation device is called analog input (AI).

Essentially, two kinds of measurement can be distinguished for the function measuring. There are sensors equipped with electronic components which can convert the captured physical values. For example, sensors that capture the relative humidity in a room belong in this category. The relative humidity can't be measured directly so the sensors are equipped with oscillating circuits with capacitors that are sensitive to moisture. This signal is then converted into an electrical value that is equivalent to relative humidity. The DDC automation device processes output signals from such sensors as active signals. The device can process two active signals:

- 0/4–20 mA and
- 0/2 – 10 V

The purpose of the selectable value range restriction is wire break control. The specified measuring range 0/2 –10 V is a lot more common in building automation. Signals with a specified measuring range of 0/4–20 mA are more common in industry automation.

However, the most common measuring values in building automation are temperature values. Due to the vast number of necessary sensors and the resulting costs, this measurement value is gathered with two-wire technique as a passive signal at the DDC inputs. To do so, the automation device measures the resistance in the range of 0–2000 Ω and converts the result into the respective temperature value. The most commonly used temperature sensor in building automation is the Pt-1000-sensor. At 0 °C, the resistance value of this platinum sensor is exactly 1000 Ω. For temperatures above the freezing point, the resistance value increases by 3.85 Ω/K. For temperatures below the freezing point it decreases respectively.

For the basic function measuring the DDC automatically monitors breaches of set limit values with 2 adjustable top and 2 adjustable bottom values.

1.4.1.4 Basic Function Switching

The purpose of the basic function switching is the control of e.g. motors, fans or pumps in operational systems. The respective outputs of the DDC automation device are called binary output (BO) or digital output (DO). Essentially, two kinds of signal outputs can be distinguished for the function switching.

One way to control aggregates is by pulse command. In this case, an output signal is used for start-up and another output signal is used for shut-off. Additionally, a self-holding circuit must be included in the control cabinet.

The other way to control aggregates is by persistent command. As long as the output contact within the DDC automation device is closed, the aggregate that is to be controlled remains switched on and if the output signal de-energizes, the aggregate is switched off. Since this method saves one output contact, it is usually preferred.

For the basic function switching it is important to notice, that apart from very small aggregates, every switching command includes feedback for execution control as well as malfunctioning notifications from e.g. a motor protection device.

1.4.1.5 Basic Function Setting

The purpose of the basic function setting is the output of continuous and non-continuous signals.

The output of continuous signals is used in operational systems to control heating and cooling valves or continuously variable engines with frequency inverters. The respective outputs of the DDC automation device are called analog outputs (AO). The DDC can output the signals:

- 0/4–20 mA and
- 0/2–10 V

Actuators used in building automation primarily process the voltage signal.

A distinctive function is the output of non-continuous signals. This function is used to control e.g. multistate fans. The basic function setting uses the same kind of binary outputs (BO) as the basic function switching. To control a 2-stage fan with the stages 0–I–II by pulse command, three binary outputs are required.

1.4.2 System Information Schema

The basic functions can be assigned to the different systems by combining information from the information list described above and the system information schema (see Fig. 1.16). This schema is colloquially called control diagram. In this case, the picture shows the system information schema of a mixed air ventilation system. The information below the double line include a schematic depiction of necessary control tasks for the DDC automation device. The bottom right corner contains a depiction of the control sequence for the temperature control component CO1 that controls the room temperature. The depicted tasks will be elaborated in Sect. 1.4.3.

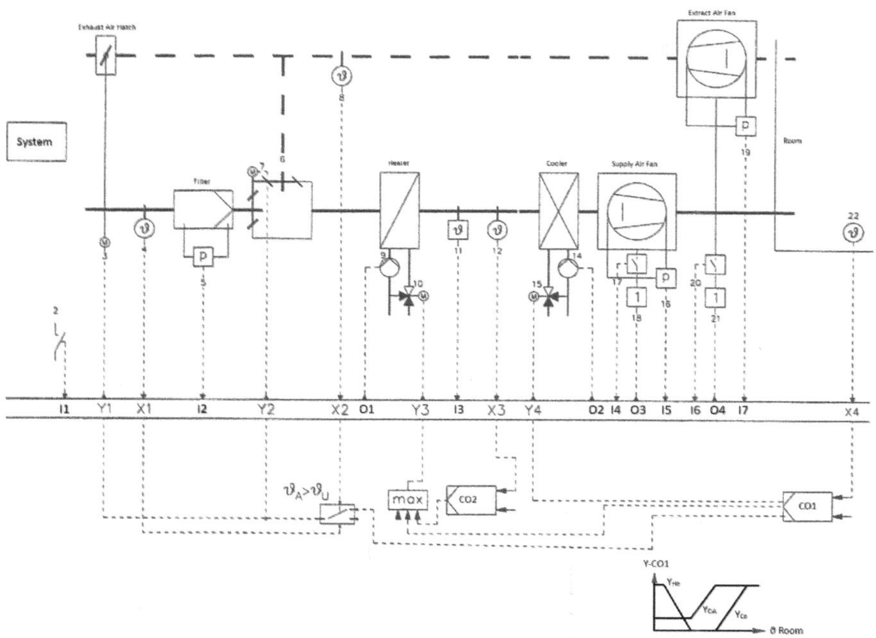

Fig. 1.16 System information schema of a mixed air ventilation system

The double line also represents the system interface or terminal block. This is where all the aggregates are connected to the DDC automation device using the basic functions explained above. The small arrows point in the direction of primary information traffic. Arrows pointing up towards the system represent output signals. Arrows pointing downwards represent input signals for the DDC automation device. The letters on the double line are abbreviations for input and output functions commonly used in control systems engineering:

- Reports are abbreviated with the letter I,
- Measurement values are abbreviated with the letter X,
- Setting signals are abbreviated with the letter Y and
- Switching commands are abbreviated with the letter O.

The setting signal for the heating valve is called Y_{He} as seen in the control sequence at the bottom right of the schema.

To be able to assign every input and output function to the respective aggregate, all the sensors and actuators depicted in the system information schema are consecutively numbered. This corresponds to the consecutive rows in the information list. For example, number 2 identifies a reporting contact. Regarding the notation, this contact can be identified as a normally open contact. The primary direction of information traffic is from the system towards the DDC automation device. In this case it is a local or remote switch installed in the system. The purpose of this switch is to enable manual operation with simultaneous shutdown of the automatic control (e.g. for test runs).

The complete assignment of basic functions to the mixed air ventilation system can be found in Figs. 1.17 and 1.18.

The depicted information list is an Excel chart. After the assignment of all basic functions has been completed, the necessary amount of input and output functions for the DDC automation device can simply be read in the total line following row 35. In the case mentioned above, 4 binary outputs, 4 analog outputs, 15 binary inputs and 5 analog inputs are required. Additionally, the switching output by bus must be considered.

1.4.3 Functions in a Ventilation System

This section gives an example of the necessary control and regulation functions for a DDC automation device assigned to a ventilation system.

1.4.3.1 Frost Protection Monitoring

In the mixed air ventilation system depicted in Fig. 1.16, number 11 identifies a so-called frost protector. A frost detector is a special kind of temperature sensors

VDI 3814 Page 1

BA information list

1) Persistent command: e.g. 0,I,II = 2 BO
Pulse command: e.g. 0,I,II = 3 BO
Setting command: z.B. closed-0-open = 2 BO
pulse width module = 1 BO
2) active or passive
3) Only shared communicative data points from external system for interoperability functions
4) per input user address to condense, delay and suppress notifications
5) per output user address

Row	System: Ventilation System: Mixed Air Ventilation System	In-/Output-Functions										Processing Functions																									
		Physical					Communicative 3)4)					Monitoring						Controlling					Regulating								Calculating/C						
		Binary Output Switching/Setting 1)	Analog Output Setting	Binary Input Reporting	Binary Input Metering	Analog Input Measuring 2)	Output Switching	Output Setting/Setpoint	Input Reporting	Input Count Value	Input Measured Value	Setpoint Value fixed	Setpoint Value floating	Operating Hours Capture	Event Counting	Command Execution Control	Notification Processing 4)	System Control	Engine Control	Switchover 5)	Follow-up Control 5)	Security/Frost Protection Control	P-Regulation	PI/PID-Regulation	Setpoint Conduction	Setting Output constant	Setting Output 2-Point 6)	Setting Output Pulse Width Modulation	Limitation Setpoint	Parameter Switching	h,x-guided Strategy	Arithmetic Calculation 7)	Event Based Switching	Time Based Switching	Floating On/Off Switching	Cyclic Switching	Automatic Night Setback
	Designation (e.g. DP-Name with Nr.) – Section	1					2					3						4					5														6
	Column	1	2	3	4	5	1	2	3	4	5	1	2	3	4	5	6	1	2	3	4	5	1	2	3	4	5	6	7	8	1	2	3	4	5	6	7
1	System Command						1						connected by bus																								
2	Local/Remote Switch			1																																	
3	Exhaust Air Hatch		1																																		
4	Outdoor Temperature					1																															
5	Filter Protector			1																																	
6	Feedback Mixed Air Hatch					1																															
7	Outdoor Air Hatch		1																																		
8	Extract Air Temperature					1																															
9	Heater Pump	1		2									incl. Feedback / Malfunction Feedback																								
10	Heater Valve		1																																		
11	Frost Protector			1																																	
12	Preventive Frost Protection					1																															
13	not assigned																																				
14	Cooler Pump	1		2									incl. Feedback / Malfunction Feedback																								
15	Cooler Valve		1																																		
16	V-Belt Protector Supply Air Fan			1																																	
17	Repair Switch Supply Air Fan			1																																	
18	Supply Air Fan, 1-Stage	1		2									incl. Feedback / Malfunction Feedback																								
	Total Functions	3	4	11		4	1																														

Output Date YY-MM-DD		Name	inspected	Plan Creator:	Project:
Rev. 1					
Rev. 2					Springer-Verlag
Rev. 3					

Fig. 1.17 Part of an information list with assignment to a mixed air ventilation system

that measures the temperature right behind the heater by means of a capillary tube filled with liquid. Should the temperature fall below a specified limit at any point, the frost detector will send a signal to the DDC automation device. The actual measurement value is irrelevant in this case since only a normally closed contact switches its state in the frost detector.

During normal operation, the temperature behind the heater can't fall so low that the water in the heater is at risk of freezing. This means that a malfunction has occurred. To prevent the system from freezing and the following water damages, a frost protection scenario standardized for North European regions is implemented by building automation:

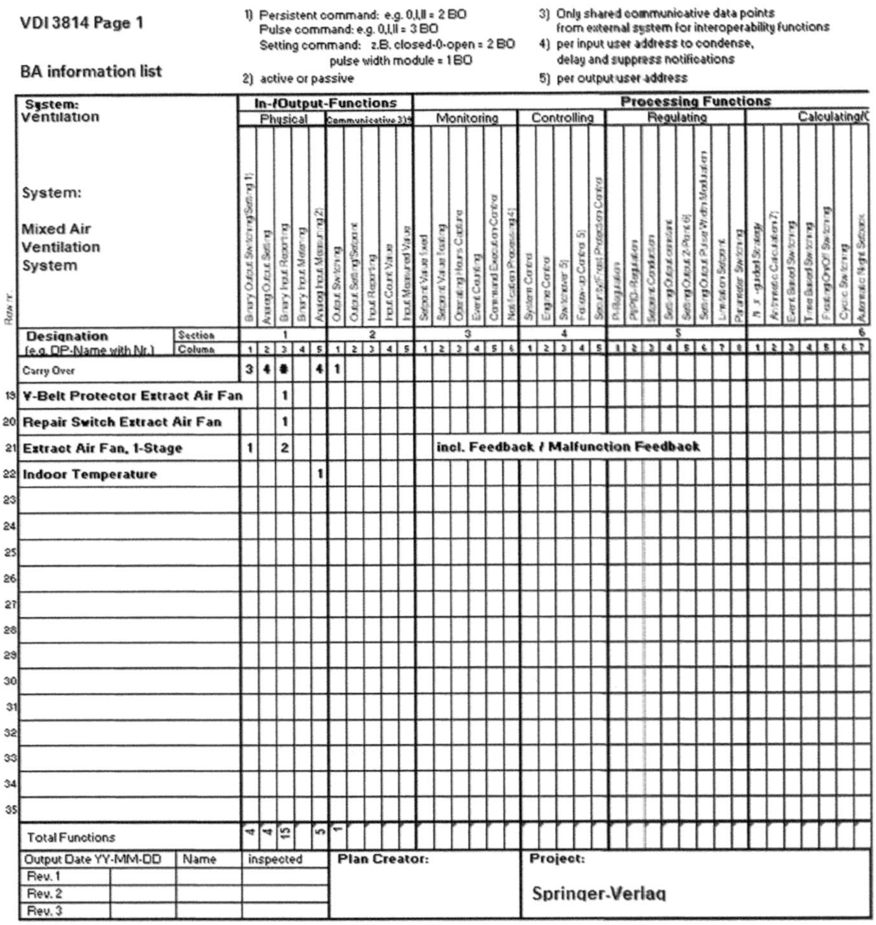

Fig. 1.18 Part of an information list with assignment to a mixed air ventilation system

- Activation of the frost detector for temperatures lower than e.g. +3 °C
- Closing the outdoor air hatch and the exhaust air hatch and completely opening the mixed air hatch
- Activation of the heater pump and completely opening the heater valve
- Deactivation of fans to prevent further indraught of cold air
- Interlocking of the entire system until acknowledgement
- Alarm notification for the control computer.

1.4.3.2 Preventive Frost Protection

A major drawback of an activation of the frost protector as seen in Sect. 1.4.3.1 is the interlocking of the system after completing the frost protection scenario. Usually, a system can only be unlocked again locally. In order to prevent this, an additional temperature sensors is installed behind the heater in some systems. This sensor is identified by number 12 in the mixed air ventilation system seen in Fig. 1.16. Its output signal is processed in component CO_2.

During normal winter operation mode, the temperature behind the heater is always above +15 °C. During automatic night setback, the temperature should never fall below this value either in order to prevent reaching the dew point with respective formation of condensate and possible mold formation in the rooms.

Therefore, a control sequence for preventive frost protection as seen in Fig. 1.19 is implemented. For temperatures below +12 °C, the component CO_2 sends a setting signal to the heater valve in order to slowly open it. This setting signal overwrites the setting signal for the heater valve sent by the main control component CO1. For canal temperatures of +6 °C or below behind the heater, the heating valve will be completely opened. Should the temperature continue to fall after this point, the frost protection monitoring as seen in Sect. 1.4.3.1 takes action.

1.4.3.3 Start-up Function for Low Outdoor Temperatures

Putting a ventilation system into service with low outdoor temperatures often leads to an activation of the frost detector resulting in interlocking of the system. In order to prevent this, a start-up function for low outdoor temperatures is implemented in the DDC automation device.

- Outdoor air hatch and exhaust air hatch completely closed, opening the mixed air hatch completely.
- Activation of heater pump and opening the heater valve completely as part of a pre-rinse-mode to preheat the heater. The duration of the pre-rinse-mode depends on the measured outdoor temperature.

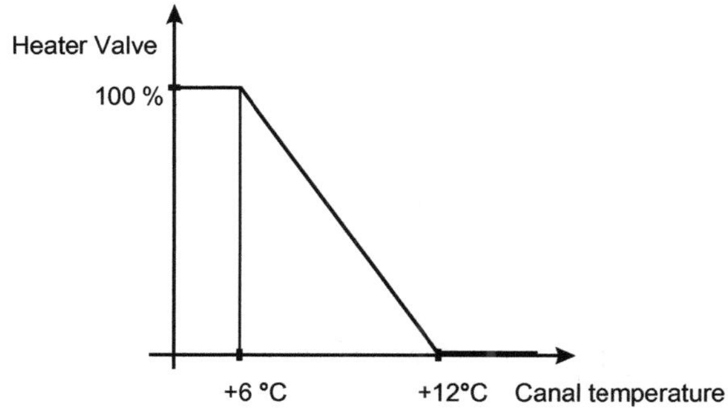

Fig. 1.19 Control sequence of preventive frost protection

- After completing pre-rinse-mode: Activation of normal operation as described in Sect. 1.4.3.4 by slightly opening outdoor air hatch and exhaust air hatch for fresh air supply and closing the mixed air hatch by the respective amount.

1.4.3.4 Control Sequence for Room Temperature Regulation

Ventilation systems in functional buildings are used to regulate the temperature in the connected rooms. With the mixed air ventilation system as seen in Fig. 1.16, the indoor temperature can either be raised or lowered. The temperature regulation mode for the DDC automation device follows the control sequence seen in Fig. 1.20.

Therefore, the set point for the indoor temperature and the measured actual indoor temperature are noted on the x-axis. For an actual indoor temperature below the set point indoor temperature, the DDC automation device sends the setting signal Y_{He} to control the heater valve. The heater valve is opened to 100% if the actual temperature differs from the set point temperature too much. If this causes the indoor temperature to rise, the heater valve is steadily closed. Should the indoor temperature be higher than desired, the DDC sends the setting signal Y_{Co} to control the cooler valve.

Additionally, the setting signal Y_{OA} for control of the outdoor air hatch is included in the control sequence. For heating operation, the outdoor air hatch is closed until a minimum of ca. 8–12%. This ensures enough fresh air supply. For the small area around the indoor temperature set point, the room is neither cooled nor heated. In this case, the indoor temperature should be kept relatively constant by modulating the amount of outdoor air and returned mixed air.

1.4.3.5 Reversion of Effect for Hatch Control During Summer Operation

For high outdoor temperatures, according to the above control sequence a setting signal to open the cooler valve up to 100% is sent. In doing so, the maximum

Fig. 1.20 Control sequence for the indoor temperature regulation

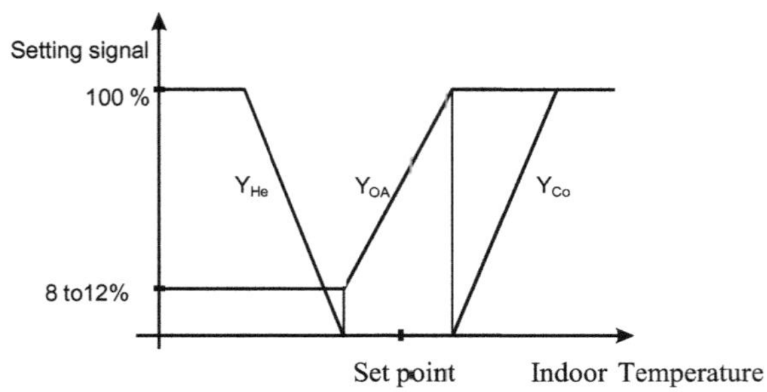

cooling power is requested and the outdoor air hatch is 100% open simultaneously. Since the exhaust air hatch is also 100% open and the mixed air hatch is completely closed, this operation mode causes the hot outdoor air to be sucked in, cooled to the desired indoor temperature and exhausted from the building afterwards. This process can be energetically improved through reversion of effect for the hatch control during summer operation.

This requires an additional logic component connected behind the output signal of main control component CO1. In Fig. 1.16, the temperature value measured by the outdoor temperature sensors is compared to the exhaust air temperature. If the outdoor temperature is above the exhaust air temperature, the setting signal for the outdoor air hatch coming from main control component CO_1 is reversed and the hatch is closed to a bare minimum. At the same time, the mixed air hatch is almost completely opened. That way, the already cooled indoor air can be reused.

1.4.4 Scope of Services and Supplies

During the construction of larger functional buildings, the installation of the different systems by the respective businesses and installment of the building automation system are usually separated. The reason behind this is to ensure, that the entire building automation is consistent and that especially cross-system energy management functions are planned and installed by one source. This is also called a turnkey construction of building automation. In this case, the supplier of the building automation system provides everything listed in Table 1.3.

In addition to all the supplies, a vast number of services is necessary until full functionality of the building automation can be guaranteed. These services can be found in Table 1.4.

Table 1.3 Scope of supplies in building automation for a turnkey construction

All sensors
All actuators
Installation incl. wiring of all sensors and actuators
Control cabinets to receive the information from the different systems
DDC automation devices
Licenses for the DDC software
Bus and network wiring to the control computer
Control computer with printer
Licenses for the control computer operating system
Licenses for the control computer management software
Furniture of the control station

Table 1.4 Scope of service in building automation for a turnkey construction

Design planning
Construction planning
Construction management
Project management
Mounting and connection of all sensors and actuators
Installation of the DDC automation devices in the control cabinets
Programming and initial operation of the DDC automation devices with data point test for all connected sensors and actuators
Generation of a visualization on the control computer with data point test for all connected sensors and actuators
Programming the energy management function on the control computer
Site acceptance test
Creating the documentation
Instruction and training of operating staff
Adjustment of the heating system at the beginning of the heating period, adjustment of the cooling system at the beginning of the summer period
Acceptance of warranty

1.5 Energy Management Functions

As well as automated control and monitoring, one of the main tasks of building automation is to ensure that the systems in a building operate efficiently. This is known as energy management. In this section, we will describe some of the energy management functions most commonly used in building automation [7].

1.5.1 Payback Period

Today efficient building automation is a standard feature of all new commercial buildings such as office blocks, hospitals or shopping centers. The reason for this becomes clear when you see how intelligent control programs can reduce a building's operating costs.

The costs of designing and constructing a building, including the building automation systems, are referred to as the total construction costs. Depending on the size of the facility, the automatic control and monitoring of the heating, ventilation, and air-conditioning systems amounts to approximately 1–1.5% of the total construction costs. For example, if the total construction costs for a multistory office building amounted to $50 million, then the building automation costs equate to approximately $625,000. Two to four percent of a building's annual operating costs can be influenced by technology. If we take 3% as the average figure from our example above, then the energy costs would amount to approximately \$1.5 million a year.

At a conservative estimate, building automation can cut a building's operating costs by about 10%. Applying this to our example, this results in a saving of ~ \$150,000 a year, which means it would take around 4 years to pay back the original investment of \$625,000. In addition to the energy saving potential, building automation also improves staff productivity and efficiency.

1.5.2 *Energy Management Functions at the Automation Level*

The programs for optimizing energy consumption can often be assigned to an individual system. In particular, if regular adjustments do not have to be made, then the necessary functions can be programmed directly into the system's DDCs. The programs can then be left to implement these functions automatically, and only need to be reprogrammed if structural changes are made to the system.

1.5.2.1 Demand-Driven Setpoint Adjustment

A common example is the weather-controlled regulation of a heating system's flow temperature (see Fig. 1.21) that uses the outside temperature to adjust the heating controller's setpoint value. When the outside temperature is low, the heating system's flow temperature is increased; and when the outside temperature is moderate, the flow temperature is reduced to the lowest possible value.

This function can also be used during the summer. If the outside temperature is very high, the room temperature setpoint can be increased without the user noticing. This "summer mode" helps to save energy by reducing the amount of air cooled by the air-conditioning system. By minimizing the difference between the outside

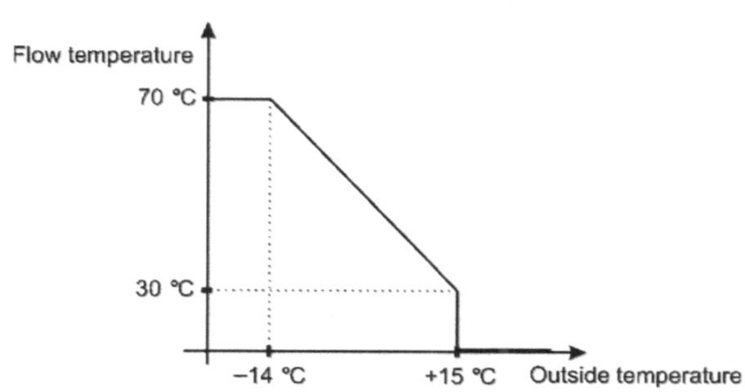

Fig. 1.21 Demand-driven setpoint adjustment

temperature and the temperature in the room, this function also prevents the user from experiencing any adverse effects to his or her health.

1.5.2.2 Enthalpy Control

Air-conditioning systems are used to control the temperature and humidity in a room. Enthalpy refers to the amount of energy contained in a substance. Enthalpy control involves calculating the optimum air flap position to ensure maximum energy efficiency, taking into account the energy content of the exhaust and outside air, as well as heating, cooling, humidification and dehumidification requirements.

1.5.2.3 Real-Time Control

There are many examples of real-time load control. A simple example uses presence detectors in a room to activate the lighting control system as soon as someone enters the room.

A more sophisticated example involves connecting all the rooms in a hotel to a booking system at the hotel reception. When a room has not been reserved, all the loads in this room remain switched off and the setpoint for the room temperature is set to the lowest acceptable level. If, however, the hotel receives a booking for this room, the setpoint temperature is then adjusted to the normal level. When the guest inserts his or her electronic key card into the card reader to enter the room, this activates the system and he or she can then adjust the temperature as desired.

1.5.2.4 Optimum Start/Stop

Optimum start/stop is a more advanced version of the scheduled start/stop program used at the management level (see Sect. 1.5.3.1). Scheduled start/stop programs are programmed to switch the system(s) on and off at specific times. The optimum start/stop program, on the other hand, is self-regulating and uses the outside and inside temperatures and the thermal characteristics of the building to calculate the optimum times to start and stop a system. In other words, the latest time a system can be switched on in the morning, and the earliest time it can be switched off in the evening (see Fig. 1.22).

1.5.2.5 Day/Night Setback

This relatively simple energy management function is used during the summer. The idea here is to delay turning on the building's air-conditioning system in the morning as long as possible. At night, when the outdoor temperature is lower than

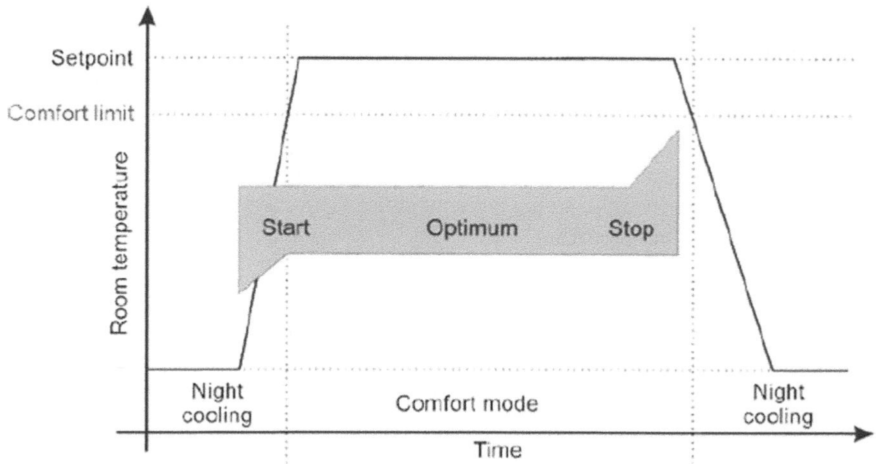

Fig. 1.22 Optimum start/stop for winter mode

the indoor temperature, all the ventilation systems only circulate outside air through the building. This mode continues to run until the early hours of the morning. The building and rooms store the cool energy.

1.5.2.6 Zero-Energy Range Control

Public buildings such as theatres, cinemas, or shopping centers, in particular, are designed to meet the requirements of the average user, and therefore often have HVAC systems for heating, cooling and ventilating rooms or other public areas.

Energy costs can be reduced by defining a temperature range, within which the system neither cools nor heats the air in the building, but simply regulates the position of the flaps to either re-circulate the internal air or to draw in air from the outside (zero-energy band, see Fig. 1.23). The disadvantage of this function is that

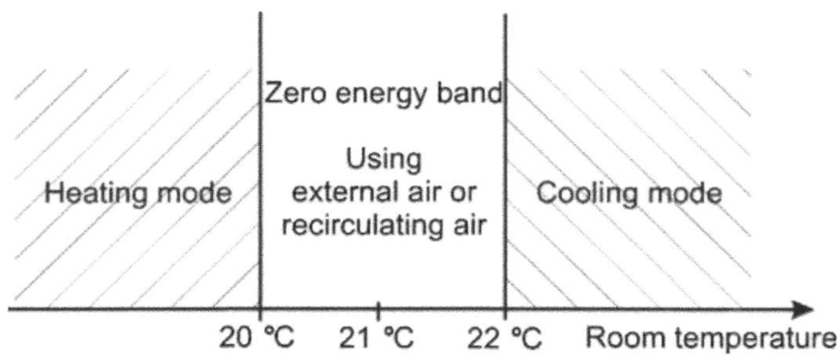

Fig. 1.23 Zero-energy range control

it is relatively difficult to accurately regulate, because you cannot accurately adjust the temperature to a predetermined setpoint.

Zero-energy range control was originally developed for use with analog controllers. Due to their significant control deviation (up to +1 K or −1 K), the analog heating controller's characteristic curve would often overlap with that of the cooling controller, resulting in both heating and cooling modes being activated at the same time to achieve the setpoint temperature. The zero-energy range prevents this phenomenon from occurring.

1.5.2.7 Duty Cycling

Another way of saving energy is regularly shut down some of the large loads (Fig. 1.24). Duty cycling is hard to control, but works well in oversized installations.

1.5.3 Energy Management Functions at the Management Level

If you need to constantly adjust a program to building's operating time or mode, then it is better to run the energy management programs from a control computer. This means you can make adjustments using an easy-to-use menu-driven user interface.

1.5.3.1 Scheduled Start/Stop

Another energy management function involves programming scheduled processes. Energy costs can be reduced by turning off non-essential systems when the building is not in use.

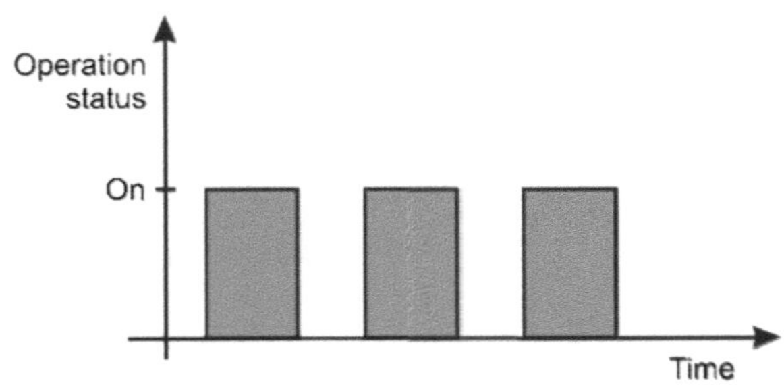

Fig. 1.24 Duty cycling

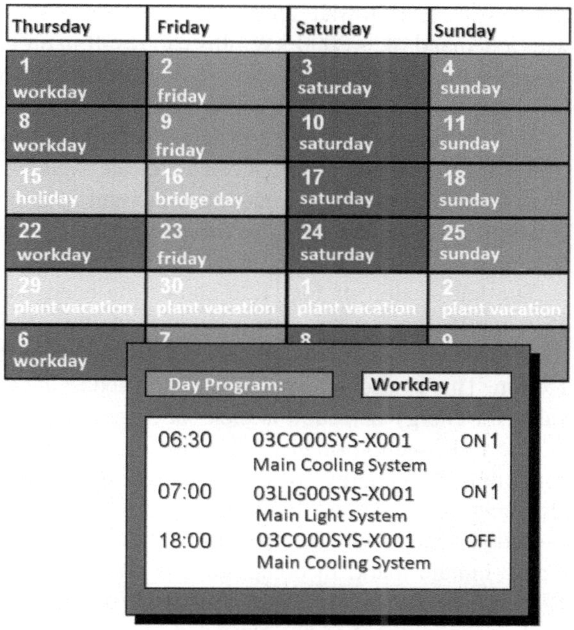

Fig. 1.25 Scheduled Start/Stop

A lighting control timer-switch program can be programmed to turn on certain lights at specific times depending on the time the sun rises, for example, when the caretaker needs to unlock buildings on campus during his morning rounds.

Figure 1.25 shows the interface of a control computer for an administration building. Different macros can be stored as day programs for different day types. Typically, the operating staff assigns these to the year conditions at the start of a new year.

These programs are also often used in private residential buildings. All heating controllers have timer-switch programs for reducing the temperature at night. In commercial buildings, these functions are usually programmed into the main control computer, which means you can easily make changes as and when you need to. For example, you can make one-off adjustments to the timer-switch programs in the main control computer for evening conferences or events.

1.5.3.2 Limiting Peak Load

In many countries the customer and energy supplier agree a rate for power demand and energy consumption. The power demand rate is paid annually and covers the amount of power made available based on the building's electrical connected load. Prices can vary depending on the contract and the amount of energy consumed—but the greater the power demand the higher the price. For example, if the demand charge is \$200 (kilowatt annum), then for an office building with 2000 employees with a demand of ~2000 kW, this can result in an annual five-figure sum. Energy consumption, on the other hand, is set at a price per consumed kilowatt hour and is considerably lower than the power demand rate.

The limiting peak demand program records the amount of electricity consumed over a 15-min period, and then calculates the average power demand for this period. The program then estimates the peak amount of power demand during this period. If the program notices that the demand is about to exceed the value agreed in the contract, it shuts off selected large loads (Fig. 1.18).

The left section of Fig. 1.26 shows the energy consumption over a period of 15 min. If the energy consumption in this time period does not exceed E = 500 kWh, it is equivalent to receiving an average of P = 2000 kW power over 15 min. This is also the limit P_{MAX} specified in the contract in this case. The count value for energy detection is then set to zero again.

The right section of Fig. 1.26 shows a higher power demand due to operating conditions. The energy consumption over the period of 15 min is predicted as too high and the specified limit of P_{MAX} = 2000 kW over 15 min will hence be exceeded. That is why the program steps in and starts to shut off large loads around the 10-minute mark. In doing so, it is ensured that the energy consumption over 15 min does not exceed E_{MAX} = 500 kWh and that thereby the limit P_{MAX} specified in the contract is not exceeded either. After the 15-minute time period, the large loads will be powered up again.

This program helps to prevent the customer from having to pay a premium price for extra electricity, and also enables the customer to monitor the average amount of power used and to reduce this amount by optimizing in-house processes. These results can then be used when negotiating the contract with the energy supplier for the following year.

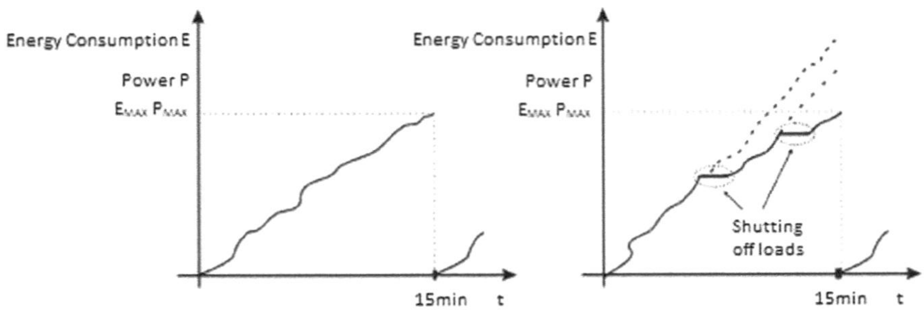

Fig. 1.26 Limiting peak load

1.5.3.3 Monitoring Energy Consumption

The energy costs in many buildings are neither recorded nor calculated. Significant savings, however, can be made, particularly in research or production facilities, if you can determine exactly how much energy each load is using (see Fig. 1.27).

Many of us usually compare our most recent household energy bills with those from the previous year. This enables us to see how much energy we use, how much it costs, and where we can make savings. In doing so we can save up to 10% a year.

This idea can also be easily and affordably implemented in commercial buildings by combining the meter readings from all the types of energy and media. Furthermore, by connecting the control computer to a commercial accounting system, you can then create a report automatically.

By combining these energy monitoring measures with in-house energy consultancy services, you can reduce energy costs even more.

1.5.3.4 Energy Saving Contracting

Another form of energy control is the energy saving contracting or simply performance contracting. During performance contracting, a rough analysis based on building specific energy values, operating modes of large loads and energy supply contracts is used to determine potential energy savings.

If enough energy saving potential is determined, a detailed analysis of actual energy savings is carried out by installing energy saving technology. These analyses also include the necessary capital needs as well as the payback period by

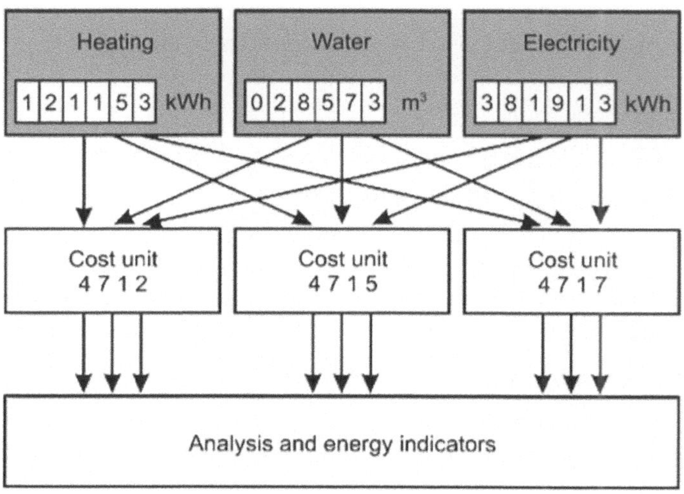

Fig. 1.27 Monitoring energy consumption

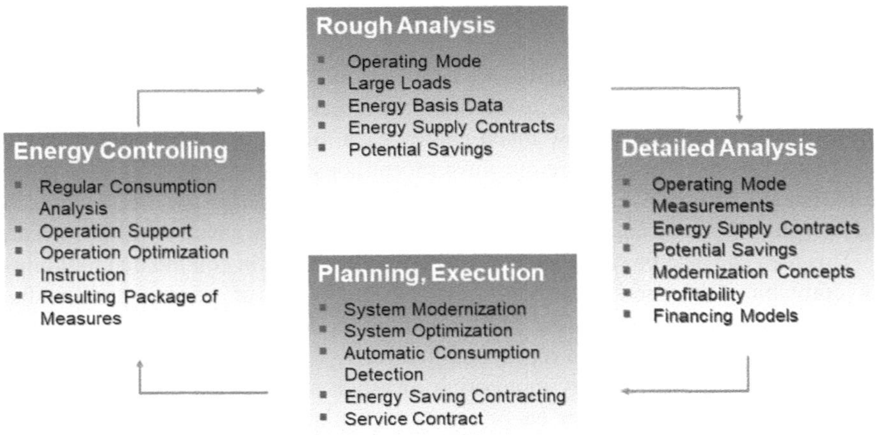

Fig. 1.28 Energy saving contracting

investment backflow in the form of saved energy costs. They are the technical and economical basis for implementing the respective measures (Fig. 1.28).

There is often no budget for installing modern energy saving technology, especially in the public sector. In this case, the service provider delivers in advance and installs the technology that is necessary to achieve the previously calculated savings. The reflux of capital in a period of 5–8 years through saving energy costs is assured by contract.

During this period, both contracting parties benefit from the implementation of energy saving measures. The service provider continually monitors the correctness of the implemented measures during performance contracting. After the contracting period, the ownership of the system usually transfers to the customer.

1.6 Comfort, Convenience and Energy Management Functions in Room Automation

Room automation is becoming increasingly important in both residential and commercial buildings. As well as the increase in comfort and convenience and the array of potential energy management functions, commercial buildings also need to have a high level of flexibility. You need to be able to easily customize these buildings to meet any change in use or requirements, for example, converting a conference room into two separate offices for new employees. By implementing well-planned building control, you can adjust many programs to meet the specifications of each of the new rooms. Traditionally this would have meant having to completely rewire the electrical installation. Instead you now simply reprogram the necessary components.

In private residential buildings, on the other hand, security is far more important than flexibility. After all it is not often we need to convert a bedroom into a kitchen.

Below is a summary of the comfort and convenience functions for each system. Some functions can also incorporate more than one system.

Heating, Cooling, and Ventilation:

- Configuration-dependent setpoint adjustment of the room temperature via an occupancy Switch or presence detector
- Use-dependent setpoint adjustment of the room temperature in hotel rooms by connecting each room to the booking system
- Adjusting the room temperature using a setpoint adjuster
- Automatically increasing the room temperature setpoint in summer when the outside temperature is high
- Shutting down the heating and cooling systems as soon as a window is opened
- Adjusting the ventilation depending on the air quality in the room.

Lighting Control:

- Configuration-dependent activation of the lighting in a room via an occupancy switch or presence detector
- Use-dependent activation of the lighting in hotel rooms by connecting each room to the booking system
- Using a brightness sensor to constantly control the lighting in a room
- Adjusting the brightness in the room depending on the light outside
- Diffused lighting control by adjusting the angle of the blind slats depending on the position of the sun
- Light switch for cleaners—which temporarily increases the amount of light in room
- Lighting scenes.

Blind and Shading Control:

- Controlling the amount of sunlight that is let into the room, depending on the time of day
- Diffused lighting control to prevent too much sunlight from shining directly into the room by altering the angle of the slats on a blind
- Winter/summer mode:
  - To prevent the room from heating up too much in the summer
  - To ensure that as much sunlight is let into the room as possible during the winter
- Raise the exterior blinds automatically if it is too windy outside.

Security:

- Illuminating the emergency escape route in case of a fire

- Smoke detectors in the room to control the extraction of smoke via electrically operated windows
- Displaying the building's evacuation plan on every office computer screen in an emergency
- Panic switch for switching on all the lights in the building
- Presence simulation for the lighting control system
- Room access using a key card or biometric scan.

Multimedia:

- Activating lighting scenes, for example, switching on a projector for a presentation will automatically dim the lights and close the blinds
- Customizing room settings by connecting audio and video servers
- Control the systems in a room using a touch pad, cell phone or computer.

1.7 Standardized Bus Systems and Networks in Building Automation

So that the various systems in a building can communicate with each other, they must be connected over communication pathways. In a medium-sized administrative building alone more than a thousand messages are transferred between the automation stations of the various systems and the control computer. Over the years bus systems (see Sect. 2.1.1) have become established as efficient and economical solutions.

For data transmission manufacturers of building automation systems originally used proprietary solutions specially developed for process automation. The main task here was to integrate the operational systems controlled by the manufacturer's own DDCs (see the systems shown in Table 1.1).

Over the years the demands on building automation have increased considerably. In the past you also needed to be able to integrate systems that were fitted with controllers from different manufacturers. To install a burglar alarm system you would have needed to connect two different bus systems so that they could communicate with each other. This would have meant that one of the manufacturers would have had to reveal its protocol to the other, which would not have been in the manufacturer's best interests. As a result, and due to the increasingly complex nature of these systems, the demand for open-protocol bus systems has risen dramatically.

Another reason for the increasing demand for open bus systems was the dependency of the customer on individual suppliers. A customer could not use components from different manufacturers in the same installation. If at a later date, the customer needed to make changes or additions to this installation, then he or she would have to buy all additional components from the same manufacturer [6].

1.7.1 Bus System and Network Requirements

Since the beginning of the 1990s task forces from various companies, as well as operators and planning engineers, have succeeded in establishing open bus systems on the market. There are still, however, a number of proprietary solutions out there.

If you look at the main areas of use and hierarchical structure of building automation shown in Fig. 1.29, you will notice that bus systems are used for different functions.

Building control components used in lighting systems only need to carry out relatively simple tasks, for example, sending and receiving switching commands within the confines of a room. Whereas heating and ventilation systems use DDCs that process and transmit switching commands as well as signals for transmitting measured values. The demands are therefore much higher than those placed on the building control bus systems.

At the management level, you need to couple these individual systems with other computer systems and programs to execute high-level energy management functions such as timer-switch or limiting peak load programs.

1.7.2 Bus Systems and Networks: Areas of Use

1.7.2.1 Lighting and Shading Control Using Konnex

The installation of lighting control and electrically operated shading systems is usually carried out by electrical installation specialists. To ensure the long-term future of this trade, many European companies have developed system components that enable the automatic implementation of comfort and convenience functions.

In 1990, leading manufacturers of electrical installation technology joined forces (forming the European Installation Bus Association (EIBA)) to develop a bus

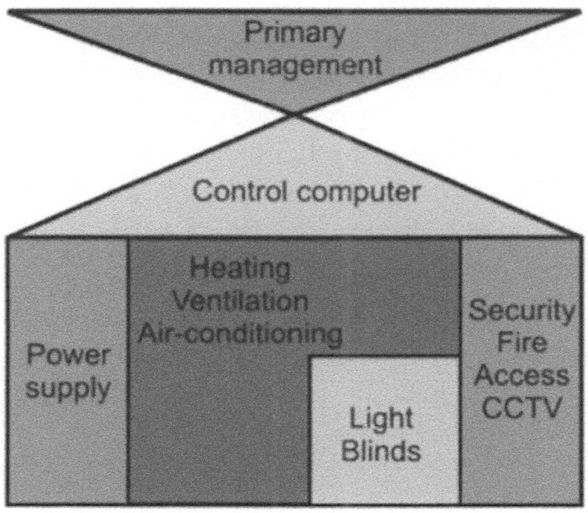

Fig. 1.29 Areas of use of bus systems and networks in a building

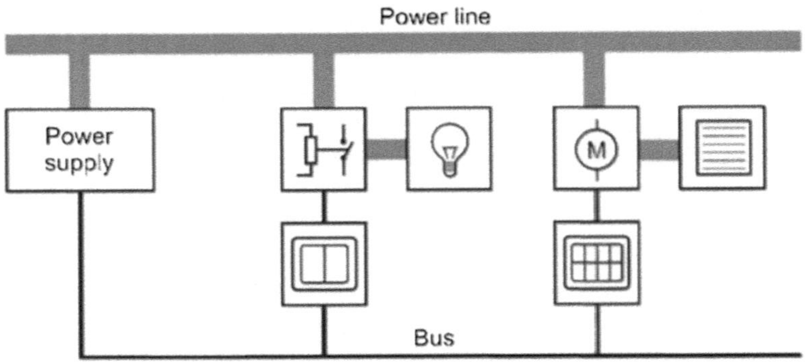

Fig. 1.30 KNX used for lighting control

system that would meet these requirements—the European Installation Bus (EIB). Since then other European organizations have joined the association—now known as the KONNEX Association. EIB is now referred to as KNX (see Chap. 3). KNX is a standardized bus system that enables data transmission between various devices and systems from different manufacturers [9].

KNX implements all necessary building control functions and can be programmed and commissioned by qualified electricians. The transmission rate is low, but sufficient for the transmission of switching and controlling commands.

KNX was not developed for controlling operational systems and, as a result, cannot be used for transmitting many analog signals. Nevertheless, it has still been very successful in Europe because it is easy to install and maintain (Fig. 1.30).

1.7.2.2 Using LonWorks to Control Heating, Ventilation and Air-Conditioning Systems

To control operational systems, a variety of measured values, setpoints, and other parameters need to be processed. The software applications place higher demands on the processor and the software engineer. The LonWorks technology (LON) was developed to meet these demands (Fig. 1.31).

LON is a universal system for automation (see Chap. 4) developed by the American company Echelon. LonWorks' scope means it can be used in centralized DDCs and in decentralized building control components. The LonMark Interoperability Association certification guarantees the compatibility of devices from different manufacturers [8].

In the field of lighting control, LON is KNX's main competitor in Europe. If you want to connect room automation to the operational systems over the same bus system, then LON has a clear advantage over KNX.

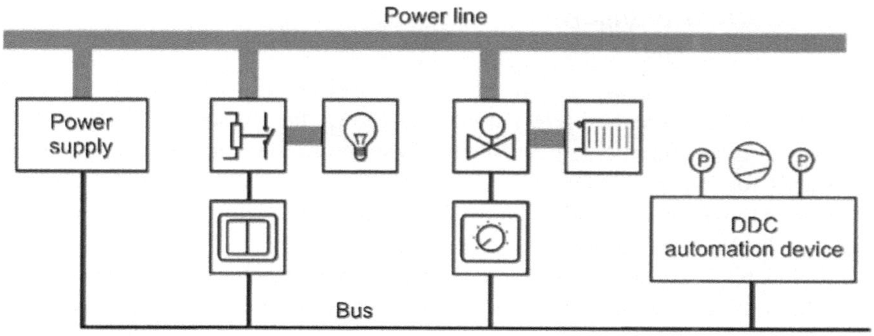

Fig. 1.31 System structure of LonWorks

1.7.2.3 Connecting Control Computers via BACnet

Extensive building automation systems in hospitals, universities or administration buildings, often have several operating stations. If you wish to extend the building by constructing an additional wing that also has an extensive building automation system, you would need to connect these systems to the existing control computer (Fig. 1.32). To do this, American building control engineers developed a standardized communication system called Building Automation and Control Network (BACnet).

BACnet (see Chap. 5) has an object-oriented structure with a wide range of functions. BACnet is tailor-made for building applications and can be used at all level of building automation (see Fig. 1.7). An efficient high-capacity network such as BACnet is not used in building control systems because it is too expensive.

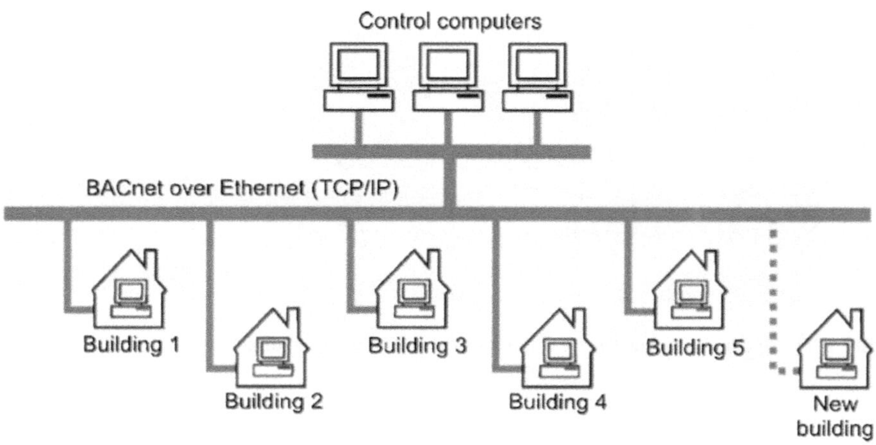

Fig. 1.32 Data exchange between buildings

1.7.3 Current Standards

There are many proprietary bus systems and networks on the market, but only a few have actually been standardized. Figure 1.33 shows the standards in 2015.

The European and international standardization processes are drawn out affairs. EN stands for European Standards and ISO/IEC refers to International Standards from the International Organization for Standardization and from the International Electrotechnical Commission.

LON technology and the Konnex, BATIbus, and EHS bus systems are used at the field level and are included in the international standard EN ISO/IEC 14908, in the US ANSI/ASHRAE 135, in the European standard EN 50090 and in the international standard ISO/IEC 14543. BATIbus is widely used in southern Europe and the Electronic Home System (EHS) is a bus system for transmitting data using the power supply network.

BACnet (EN ISO 16484-5) is the main system used at the automation level. Other protocols used at this level include Profibus (used in industry automation) and the French Factory Instrumentation Protocol (FIP or WorldFIP). Here LON is not a standardized automation level protocol, but it is as a transmission medium for BACnet.

BACnet is the only protocol used for data communication at the management level. In Europe, it has replaced the *Firmenneutraler Datenaustausch* (FND) protocol, which was developed by public authorities to enable data exchange between control computers made by different manufacturers.

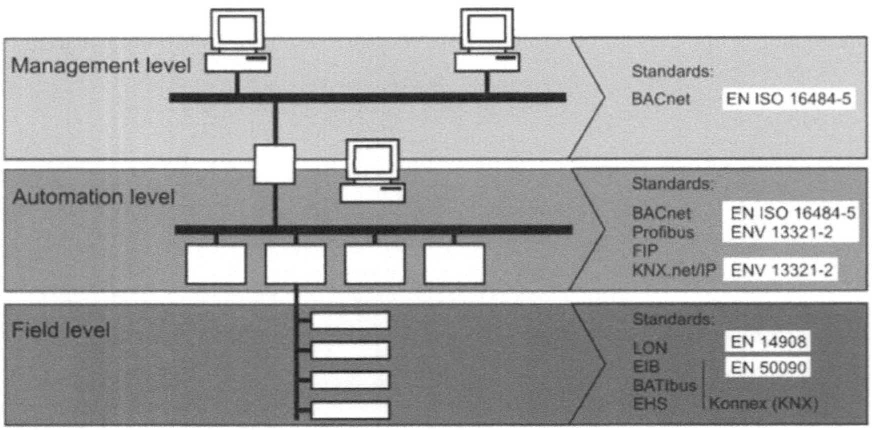

Fig. 1.33 Standardized bus systems and networks in building automation

1.8 Exercises

Exercise 1.1
Explain the difference between building automation and building control.

Exercise 1.2
What is the system interface in control cabinets?

Exercise 1.3
Why are most commercial buildings nowadays equipped with building automation?

Exercise 1.4
Name the 5 basic functions in building automation. What are their purposes?

Exercise 1.5
What is the importance of system information schemas for building automation?

Exercise 1.6
One function for controlling ventilation systems is the reversion of effect for hatch control during summer operation. Explain this function and its purpose.

Exercise 1.7
Is "Limiting Peak Load" an energy saving function?

Exercise 1.8
Give examples of energy management functions that can be used in hotel rooms.

Exercise 1.9
What kind of comfort and convenience functions can be implemented in a private residential building?

Exercise 1.10
What are the advantages and disadvantages of using standardized bus systems and networks in building automation?

References

1. ABB Mannheim, Germany
2. Balow J (2012) Systeme der Gebäudeautomation. cci Dialog GmbH, Karlsruhe
3. ELKA Elektronik, Insta GmbH Lüdenscheid Germany
4. Fachverband Installationsgeräte und -systeme (1997) Handbuch Gebäudesystemtechnik. Zentralverband Elektrotechnik- und Elektronikindustrie e.V., Frankfurt a. M.
5. Handbuch TAC Xenta® 280-300-401 (2002) Firma TAC
6. Hansemann T. (Hrsg.), Merz H. (2003) Kommunikationssysteme für die Gebäudeautomation—Wirtschaftlicher Bedienungskomfort in Gebäuden mit Hilfe von Bussystemen. Shaker, Aachen

7. Kranz R (1997) u. a.: Building control. Remningen-Malmsheim: expert
8. LON Nutzer Organisation e.V. (2004) LonWorks-Installationshandbuch. VDE, Berlin
9. Merz H (ed.) (2001) Kommunikationssysteme für die Gebäudeautomation—Theoretische Grundlagen und Praxisbeispiele. Shaker, Aachen
10. Staub R, Kranz HR (2001) Raumautomation im Bürogebäude. Die Bibliothek der Technik, Band 210. Moderne Industrie, Landsberg
11. VDI-Gesellschaft Technische Gebäudeausrüstung: Gebäudeautomation (GA) (2005) Düsseldorf, Blatt 1 bis 5
12. VDI-Gesellschaft Technische Gebäudeausrüstung: Gebäudeautomation—Grundlagen der Raumautomation, Richtlinie 3813 Blatt 1 bis 3. Düsseldorf, 2011 bis 2015

Chapter 2
The Basics of Industrial Communication Technology

2.1 Industrial Communication

Industrial communication refers to the communication between industrial automation devices, as opposed to the use of language that we use to communicate with each other.

The devices in an automated system or process must be able to communicate with each other. This flow of information can be classified into levels using, for example, the three-tier model used in building automation, comprising the management, automation and field levels (see Fig. 2.1 and Chap. 1). To perform the automated functions at the various levels, information is constantly exchanged within each level (horizontal communication) and between the different levels (vertical communication).

Industrial communication systems such as field buses and networks are used for horizontal and vertical communication.

2.1.1 Field Bus Communication

At the field level (on site, in the system) you will find sensors and actuators—the so-called field devices. Functions typically carried out at the field level include switching, setting, reporting, measuring and metering. Field devices used on the field bus have in-built microcontrollers and are, therefore, referred to as "intelligent" components. They send and receive messages (a series of bits) over a bus, communicating either directly with each other or with control and regulation devices at the automation level above.

A field bus is a digital serial data bus that enables communication between industrial automation devices such as meters, regulators and programmable logic controllers [IEC 61158, IEC 61784].

H. Merz et al., *Building Automation*, Signals and Communication Technology,
https://doi.org/10.1007/978-3-319-73223-7_2

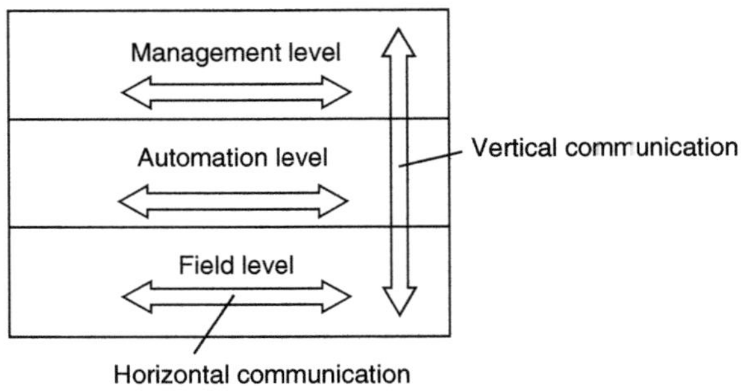

Fig. 2.1 Horizontal and vertical communication in the three-tier building automation model

Table 2.1 Examples of field buses and the fields they are used in

Field buses	Field of use
Controller area network (CAN)	Automobile engineering
Local interconnect network (LIN)	
Process field bus (Profibus)	Process and factory automation
Interbus	
Konnex (KNX)	Building automation
Local operating network (LON)	
Local control network (LCN)	
Serial real time communication system (SERCOS) interface	Drive engineering

This book focuses on KNX (see Chap. 3) and LON (see Chap. 4)

Field bus technology was first developed in the 1980s as part of the drive to increasingly decentralize automation solutions. The aim was to replace standard parallel transmission technology using analog data transfer with serial transmission technology using digital data transfer. There are currently various field bus systems on the market [1]. Their specifications (e.g., transfer rate, cable length or the number of devices) vary depending on the requirements of the particular field they are used in. Field buses can only transfer small amounts of digital data (bits and bytes) in a short space of time (μs, ms). Table 2.1 shows a selection of field buses and the fields they are used in.

2.1.2 Communication over Networks

Information is transferred from the automation level to, for example, visualization and production planning systems at the management level. Larger amounts of information are transferred than at the field level and more time is available to do so. Communication at the automation level takes place primarily over local area networks (LANs).

A network uses wires or radio waves to connect various technical systems such as computers and control devices so that they can communicate with each other. Communication takes places according to specific rules (protocols) that can be based on the ISO/OSI reference model.

The Building Automation Control Network (BACnet) communication protocol, developed by the American Society of Heating, Refrigeration and Air-Conditioning Engineers (ASHRAE), has become increasingly important in building automation (www.bacnet.org).

BACnet enables devices and systems used in building automation to exchange information with each other. Data transfer with BACnet can take place using different types of networks. BACnet supports LAN technologies such as Master Slave/Token Passing (MS/TP), Local Operating Network (LON), ARCNET [www.arcnet.de], Ethernet, and also dial-up connections over telephone networks (see Chap. 5).

2.2 Digital Data Transfer: Important Terms and Definitions

Field buses and networks are used in automation technology for the digital transfer of data. The next section will explain a few key terms.

2.2.1 Key Terms

2.2.1.1 Bits and Bytes

A bit stands for a binary digit. It is the smallest unit used to represent a piece of information that consists of two values such as yes/no or on/off. The binary system is used to mathematically represent binary information. A bit represents the value 0 (zero bit) or 1 (one bit). The one bit could be assigned to the command for turning a light "on" and the zero bit could represent the command to turn the light "off."

We can use n bit to represent 2^n states, for example, $2^2 = 4$ states where $n = 2$ bit. Table 2.2 shows you the states that can be represented by binary numbers.

Eight bits make up one byte (also known as an octet).

According to IEC 60027-2, a bit is the SI unit (SI stands for *Le Système International d'Unités*) for the equivalent binary storage capacity M_e. The SI symbol for a binary digit is bit (this is equivalent to the SI symbol for seconds which is s), and B is the symbol for a byte. Note: byte is not an SI unit. The

Table 2.2 Examples of the information represented by binary numbers

Binary number	State
00	Tank is empty
01	Tank is half full
10	Tank is three-quarters full
11	Tank is full

equivalent binary storage capacity can be calculated from the number *n* of the possible states of memory available as: $M_e = \text{lb}\ n$, where lb is the binary logarithm.

If, for example, there is enough storage capacity to store $n = 256$ possible states, then:

$$M_e = \text{lb}\ 256 = 8\ \text{bit}\ = 1\ \text{B}.$$

Here the storage capacity $M_{\text{bit}} = 256$. In other words, there is enough memory to store 256 binary states (bits).

Please note that, according to IEC 60027-2, storage capacity is represented by one SI unit (symbol 1). In practice, however, it is usually represented as $M_{\text{bit}} = 256$ bit.

To express storage capacity such as M_{bit} or an equivalent storage capacity M_e, the SI unit bit and the unit Byte (B) can be combined with SI prefixes representing either binary or decimal multiples:

- 1 kbit (1 Kilobit) = $(10^3)^1$ bit = 10^3 bit = 1000 bit
- 1 MB (1 Megabyte) = $(10^3)^2$ B = 10^6 B = 1 000 000 B
- 1 GB (1 Gigabyte) = $(10^3)^3$ B = 10^9 B = 1 000 000 000 B
- 1 Kibit (1 Kibibit) = $(2^{10})^1$ bit = 2^{10} bit = 1024 bit
- 1 MiB (1 Mebibyte) = $(2^{10})^2$ B = 2^{20} B = 1 048 576 B
- 1 GiB (1 Gibibyte) = $(2^{10})^3$ B = 2^{30} B = 1 073 741 824 B

2.2.1.2 Binary and Hexadecimal Format

Since a representation of bit sequences in binary format can easily lead to unclear sequences (e.g. 0100 1011 0100 1110 0101 1000), hexadecimal format is preferred in this case. To convert a binary number into a hexadecimal number, 4 bits are condensed as seen in Table 2.3.

With a change of format, the sequence 0100 1011 0100 1110 0101 1000 can be expressed clearer as:

$$4B4E58_{\text{HEX}}$$

The bit sequence can also be expressed as a sequence of bytes. In this case, 4 bits of one byte are expressed as a hexadecimal number, e.g.:

$$4B_{\text{HEX}},\ 4E_{\text{HEX}},\ 58_{\text{HEX}}\ \text{or}\ 0 \times 4B, 0 \times 4E, 0 \times 58$$

Table 2.3 Binary, hexadecimal and decimal format

Binary	Hexadecimal	Decimal
0000	0	0
0001	1	1
0010	2	2
0011	3	3
0100	4	4
0101	5	5
0110	6	6
0111	7	7
1000	8	8
1001	9	9
1010	A	10
1011	B	11
1100	C	12
1101	D	13
1110	E	14
1111	F	15

2.2.1.3 Bit Rate

The binary digit rate (r_b or r_{bit}), usually referred to as *bit rate*, is defined as the reciprocal of the bit period (T_b or T_{bit}).

$$r_b = \frac{1}{T_b}$$

That means that the unit of bit rate r_b is *second* to the power of *negative one*:

$$[r_b] = \frac{1}{s} = s^{-1}$$

However, the more common expression for bit rate is bit per second (bit/s). In this case, *bit* is used as the unit instead of the implicit *one* [2]. Prefixes can be added to bit/s like e.g. kbit/s or Mbit/s.

The byte rate r_B is defined in analogy to the bit rate. It is usually expressed as byte per second (B/s). Prefixes can be added (e.g. kB/s or MB/s).

For KNX, 9600 bits are transmitted per second:

$$r_b \approx 9600 \frac{\text{bit}}{\text{s}} = 9{,}6 \frac{\text{kbit}}{\text{s}}$$

2.2.1.4 Modulation Rate

The modulation rate r_m (or u) is the reciprocal of the shortest duration in seconds of a signal element (symbol) used during line coding (see Sect. 2.2.5). If all the signal elements have the same duration, then the modulation rate equals the number of signal elements transmitted in a given time interval divided by the length of this time interval.

The SI unit of measurement for modulation rate (r_m *or* u) is baud (Bd) or symbols/second, named after the French telegraph engineer Jean-Maurice-Émile Baudot (1845–1903). It can be also combined with SI prefixes to form kBd or MBd.

2.2.1.5 Correlation Between Bit Rate and Modulation Rate

For binary signals, two signal elements are used for line coding the bit's two states (values). This means that the bit rate r_{bit} and modulation rate r_m are the same. However, this only applies to binary signals where only one bit is transmitted at a time.

You can also transmit, for example, two bits (dibit) at a time by using four different signal elements and assigning a specific combination of bits to each signal element (see Table 2.4).

For the "10" bit combination, the line coder sends the signal element 3. The bit rate is now twice the modulation rate. The relationship between the bit rate and modulation rate (where n = the number of signal elements) can be expressed using the following formula:

$$r_{\text{b}} = \text{lb}(n) \cdot r_{\text{m}} = \text{lb}(n) \cdot u$$

2.2.2 *Digital Data Transmission Systems*

A digital data transmission system is used to transfer data bits from a source to a sink (destination). The source bits contain information on the current state of a process, for example, motor 1 is on ("1"), motor 2 is off ("0"), tank 1 is full ("1"), tank 2 is empty ("0"), etc. The information supplied by the binary sensors is then sent to a programmable logic controller (PLC), where it is then saved and processed. The sink refers to the PLC's memory.

Table 2.4 Assigned combination of bits to signal elements

Bit combination (Dibit)	Signal element
00	1
01	2
10	3
11	4

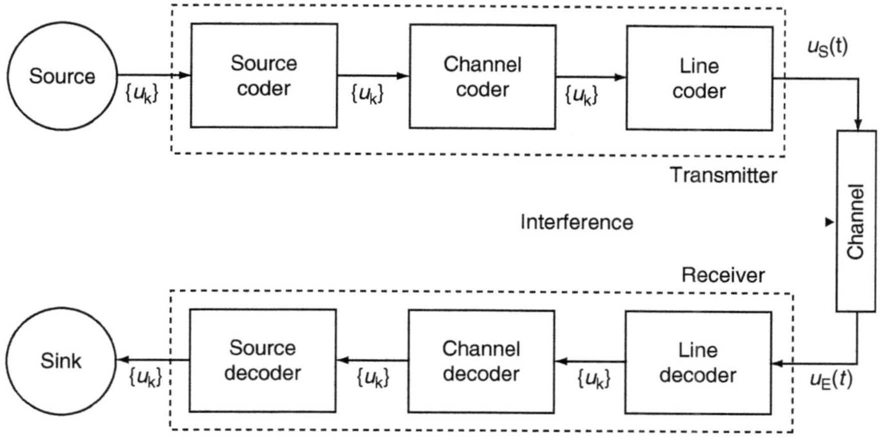

Fig. 2.2 The basic structure of a digital data transmission system

Table 2.5 The transmitter and receiver components and their respective tasks

Components	Function
Source coder	Removes redundant source data bits
Source decoder	Adds redundant source data bits
Channel coder	Adds error-correction bits with additional redundant bits
Channel decoder	Removes the error-correction bits
Line coder	Converts the bit sequence into a physical signal
Line decoder	Converts the physical signal into a bit sequence

A digital data transmission system consists of a transmitter (e.g., a sensor), a channel (e.g., a field bus) and a receiver (e.g., a PLC). Figure 2.2 shows the basic structure of such a system.

The source forwards the bits to the transmitter. The bit sequence then passes through the source, channel and line coders, before it is converted into a physical signal $u_S(t)$ and transmitted over the channel. The bits can be converted into a voltage signal for transmission over a copper or coaxial cable, a light signal over a fiber-optic cable or a radio signal using the air as a channel.

The receiver—comprising the line, channel and source decoders—reconstructs the source bit sequence from the signal $u_E(t)$ and stores it in the sink. Table 2.5 shows the functions performed by the components of the transmitter and the receiver.

2.2.3 Source Coding and Decoding

When transmitting information as a sequence of bits, as few bits as possible should be transferred at one time to reduce the transmission time. Some bit combinations occur more often than others, which means you can recode the source code words in such a way that the most frequently occurring bit combinations are assigned to the shortest code words, and those that occur the least are assigned to the longest code words. Shannon Fano Coding is a commonly used technique.

We will show you how this technique works using the following example. Supposing the source supplies four symbols: four eight-bit values A, B, C, and D (source code words):

$$\mathrm{A}:\ 01000001\quad \mathrm{B}:\ 01000010\quad \mathrm{C}:\ 01000011\quad \mathrm{D}:\ 01000100.$$

The probability (P) that each of the symbols occurs is as follows:

$$P(\mathrm{A}) = 0.5\quad P(\mathrm{B}) = 0.125\quad P(\mathrm{C}) = 0.25\quad P(\mathrm{D}) = 0.125$$

(The probability that one of the values appears is 1, in other words 100% chance).

Shannon Fano Coding:

1. Arrange the symbols in order of probability (highest to lowest).

(A)	(C)	(B)	(D)
0.5	0.25	0.125	0.125

(B and D can be arranged in any order.)

2. Divide the symbols into groups according to similar or equal probabilities.

(A)	(C)	(B)	(D)
0.5	0.25	0.125	0.125

3. Assign "0" to the left group and "1" to the right group.

(A)	(C)	(B)	(D)
0.5	0.25	0.125	0.125
0	1	1	1

4. Repeat steps 2 and 3 until each symbol has been recoded.

(A)	(C)	(B)	(D)
0.5	0.25	0.125	0.125
0	1	1	1
	0	1	1
		0	1

The symbols A to D (here values) have now been assigned the following code words:

$$A:0 \quad B:110 \quad C:10 \quad D:111$$

At first glance, it looks like nothing is gained from transforming the code, since the new code even uses 9 bits instead of 8. However, the symbol A, coded with just one bit, exists 4 times per 8 bits on average, e.g.:

ACBAACAD or CAABADAC or ADABACAC

This means, that the process still leads to an average reduction in code word length of:

$$1\frac{\text{bit}}{\text{symbol}}\cdot 0.5+3\frac{\text{bit}}{\text{symbol}}\cdot 0.125+2\frac{\text{bit}}{\text{symbol}}\cdot 0.125+3\frac{\text{bit}}{\text{symbol}}\cdot 0.125=1.75\frac{\text{bit}}{\text{symbol}}$$

Redundancies in the source code words are reduced as well. For example, the first six bits (010000) for all four measurements are identical and only the last two bits are different. These two bits contain the information that is to be transmitted. The other bits are redundant as they contain no new information.

Source coding and decoding is not generally used in field bus transmission. This is because it is often difficult to determine the probability of a source word occurring, and there are often not enough redundant source code words.

2.2.4 Channel Coding and Decoding

Errors always occur when bits are transmitted. For example, a one bit may be sent but a zero bit received, and vice versa. Data transmission is never 100% secure and errors generally occur at random. As a rule, however, the error does not occur again when the data is re-transmitted.

Channel coding rearranges the bit sequence in such a way that as few errors occur during data transmission as possible. This means that:

(a) Errors are detected by the receiver and the transmission is repeated (most common)
(b) Errors are detected and corrected

To do this additional bits are inserted into the bit sequence. This increases the number of redundant bits, but also improves error detection.

Three of the most commonly used channel-coding techniques are:

- Parity check
- Block parity check
- Cyclic Redundancy Check (CRC).

2.2.4.1 Parity Check

The parity check involves adding an extra bit—a so-called parity bit—to a bit sequence before it is transmitted. The value of this bit depends on whether

(a) The whole bit sequence including the parity bit has an even number of one bits, known as even parity (e.g., 01000011 **1**)
(b) The whole bit sequence including the parity bit has an odd number of one bits, known as odd parity (e.g., 01000011 **0**).

The transmitter and receiver must use the same type of parity check. With the parity check:

- only an odd number of bit errors is detected
- an even number of bit errors cannot be detected
- errors cannot be corrected because there is no way to determine which and how many bits have been corrupted
- once an error has been detected the data must be re-transmitted in its entirety.

2.2.4.2 Block Parity Check

The block parity check involves dividing the bit sequence into groups (or characters) of, for example, eight bits. These groups of bits are then arranged into a grid (block) and then each row and column is secured with a parity bit.

Example:

The bit sequence 10101110 01011101 11000000 01101000 is to be transmitted and secured using block parity.

Firstly, the sequence of bits is divided into four characters, each containing eight bits (see Table 2.6).

Secondly, an eight-bit check character is generated from the four original characters. This is achieved by inserting a parity bit at the bottom of each column using either even or odd parity, but not both. In Table 2.6 we have used even parity. The check character comprises the parity bits from all the columns. Now the parity bits are generated for all the characters, including the check character, again using

Table 2.6 Example of block parity

1	0	1	0	1	1	1	0	→	*1*
0	1	0	1	1	1	0	1	→	*1*
1	1	0	0	0	0	0	0	→	*0*
0	1	1	0	1	0	0	0	→	*1*
↓	↓	↓	↓	↓	↓	↓	↓		
0	1	0	1	1	0	1	1	→	*1*

either even or odd parity, but not both. The example in Table 2.6 uses even parity. Finally, the following bit sequence is sent:

10101110 *1* 01011101 *1* 11000000 *0* 01101000 *1* 01011011 *1*

45 bits are sent instead of 32 bits.

The transmitter and receiver must use the same parity check (odd or even) for the columns and rows. Block parity displays the following error detection and correction attributes:

- Detects and corrects one-bit errors
- Detects but cannot correct two and three-bit errors
- Can only detect four-bit errors if the error does not "occur in the corners of a square"
- To correct an error once it has been detected, the data must be transmitted again

Block parity is typically used in KNX.

2.2.4.3 Cyclic Redundancy Check

The cyclic redundancy check (CRC) is an error detection technique often used in field buses. If an error is detected the data is transmitted again.

In CRC a sequence of redundancy bits (frame check sequence) is appended to the bit sequence to be transmitted, forming a data frame that is exactly divisible by a predetermined binary sequence (CRC polynomial or divisor). At the destination the incoming data is then divided by this predetermined binary sequence. If there is no remainder, then the data is deemed as intact and is accepted. Even though this technique detects far more errors than the parity and block parity checks, it still cannot detect all errors.

Here is an explanation of CRC using the decimal system. For the number 57, the number 23 is the "check number" used to determine whether the two numbers have been corrupted during transmission. To do this a third number x is appended to the number 57 so that the whole number is exactly divisible by 23. In this case, x is 5 because 575/23 = 25 remainder 0. The number 575 is now transmitted. If this number is corrupted during transmission, for example, 471 is the number received, then the receiver recognizes that 471/23 = 20 remainder 11. The only drawback of the technique is that it cannot detect errors if the corrupted number is also divisible by 23.

Now we will explain how CRC works using the binary system.

We will start with a sequence of bits (frame) that is to be transmitted and a (standardized) divisor, for example, the CRC-16 check polynomial 1 1000 0000 0000 0111. The divisor is usually represented by the polynomial x such as the CRC-16 check polynomial:

$$x^{16} + x^{15} + x^2 + x^1 + x^0 = x^{16} + x^{15} + x^2 + x + 1.$$

The degree k of the check polynomial, which is required to implement the procedure, can be easily determined. For the CRC-16 check polynomial $k = 16$.

We now require a frame check sequence that when appended to the data bit sequence results in a bit sequence (data + check-bit sequence) that is exactly divided by the CRC polynomial. To determine the frame check sequence, k 0-bits are added to the data bit sequence, which is then divided by the check polynomial. The resulting remainder is the frame check sequence we are looking for. It is k bits in length.

The receiver divides the data it receives by the frame check sequence and assumes, if the remainder is 0, that an error has not occurred. This technique cannot detect errors that corrupt the data bit sequence in such a way that a multiple of the check polynomial is created.

The rules of Modulo 2 arithmetic apply when subtracting during polynomial division (see Table 2.7).

$$1 - 0 = 1,\ \ 1 - 1 = 0,\ \ 0 - 1 = 1$$

Remainders are not carried (see Table 2.8)

Worked example:

To simplify the following worked example, we will use a short (fictitious) CRC sequence: 10011 (as polynomial: $x^4 + x + 1$). The data bit sequence is as follows:

11 0101 1011

1. Add k 0-bits to the data bit sequence.
   The degree of the polynomial is $k = 4$, in other words, four zero-bits are added:

   11 0101 1011 0000

Table 2.7 Subtraction rules in binary system according to modulo 2

1	1	0	0
−0	−1	−0	−1
1	0	0	1

Table 2.8 Binary subtraction and modulo 2 arithmetic subtraction

Binary subtraction (carry-over)						Subtraction according to modulo 2 arithmetic (no carry-over)					
	1	1	0	1	1		1	1	0	1	1
−	1	0	1	0	0	−	1	0	1	0	0
	0	0	1	1	1		0	1	1	1	1

2. Divide the data bit sequence (including the 0-bits) by the check polynomial.

```
1101011011 0000 : 10011 = 1100001010
-10011
------
 10011
-10011
------
      10110
     -10011
     ------
        10100
       -10011
       ------
         1110 Remainder
```

The resultant check-bit sequence is: 1110.

3. Replace the k zero-bits with the frame check sequence and send the whole bit sequence: 11 0101 1011 1110.
4. The receiver divides the received bit sequence by the check polynomial and assumes, if the remainder is 0, that an error has not occurred.

2.2.5 Line Coding and Decoding

The outgoing bit sequence from the channel coder needs to be converted into a physical signal that can be transmitted over the channel. Copper cable is often used as the transmission medium for field buses. The bit sequence is converted into a voltage signal. This is usually a binary signal consisting of two signal elements: the first represents the zero bit and the second represents the one bit. The total number of signal elements contained in the signal corresponds to the number of bits in the bit sequence.

There are many factors to consider when defining signal elements such as the available bandwidth on the channel, whether the signal is dependent on a DC component or whether the clock signal can be recovered from the signal.

There are a variety of line codes to choose from. The most widely used are the Non-Return-to-Zero (NRZ) code and the two Manchester codes (Biphase-L code and the Differential Manchester code).

2.2.5.1 The Non-return-to-Zero Code (NRZ)

The NRZ code uses signal elements with the constant levels U_H and U_L with T representing the duration (Fig. 2.3).

The signal elements can be arbitrarily assigned to the logical states "0" and "1". For the RS-232 serial interface, the signal elements can be assigned as follows:

$$\text{“0“} \rightarrow U_H \text{ with } 3\,\text{V} \leq U_H \leq 15\,\text{V and “1''} \rightarrow U_L \text{ with } -3\,\text{V} \leq U_L \leq -15\,\text{V}.$$

The line coder then converts the bit sequence 010010 into the following signal (Fig. 2.4).

NRZ coded signals are generally bound to the DC component. If longer streams of zeros or ones are transmitted, the receiver may not be able to determine the unit interval *T*, because the signal no longer "changes states," and instead just stays at one value. The Manchester code does not have this problem.

KNX uses signal elements similar to those used in the NRZ code (see Chap. 3).

2.2.5.2 The Manchester Code (Biphase-L)

The Manchester code (Biphase-L) uses two signal elements with *T* representing the duration. After *T/2* there is a transition from level U_H to $-U_H$ or vice versa (Fig. 2.5).

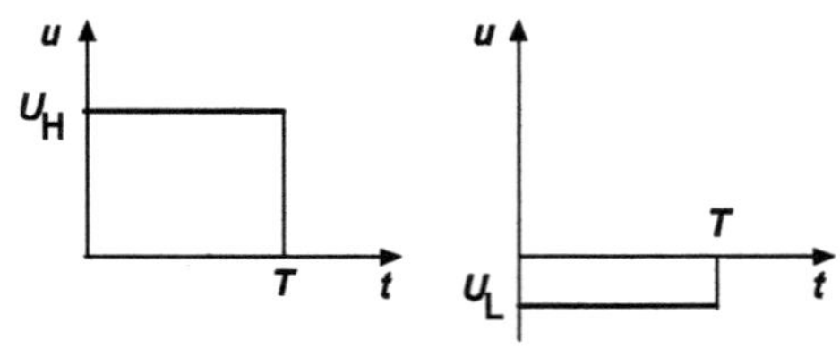

Fig. 2.3 NRZ code signal elements

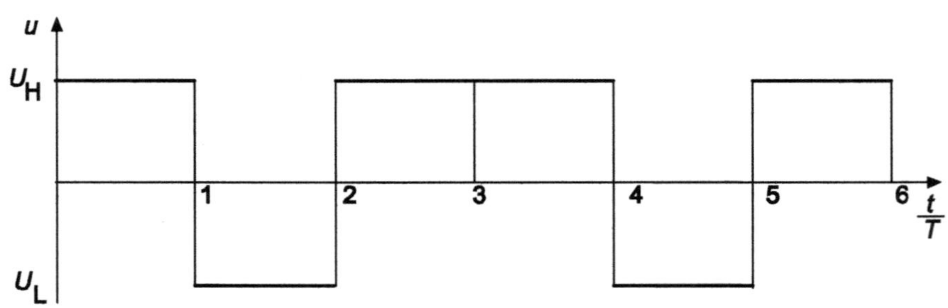

Fig. 2.4 An NRZ coded RS-232 voltage signal

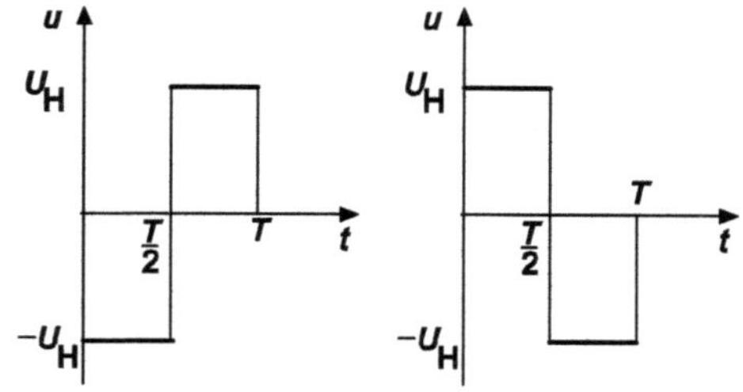

Fig. 2.5 Signal elements in the Manchester code

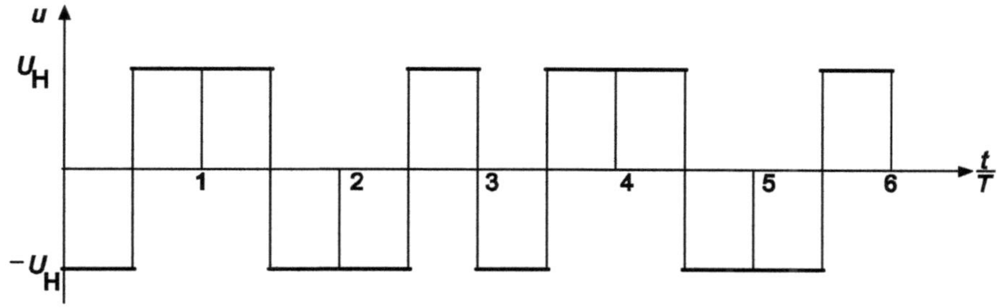

Fig. 2.6 Example of the Manchester code (Biphase-L)

The signal elements are assigned to the logical states "0" and "1" as follows:

- The "0" is assigned to the ascending transition in the middle of the interval
- The "1" is assigned to the descending transition in the middle of the interval

They can also be assigned in the other direction.

The line coder then converts the bit sequence 010010 into the Manchester signal in Fig. 2.6.

The advantages of the Manchester code are that it is not bound to the DC component and it contains sufficient clocking information: A transition occurs after the unit interval *T* at the latest. The Manchester code requires double the amount of bandwidth for the same unit interval *T* compared with the NRZ code because the rectangular impulses have double the frequency.

The Manchester code is an important part of data transmission in local Ethernet-based networks (see Chap. 5).

2.2.5.3 The Differential Manchester Code

As with the Biphase Manchester code, the Differential Manchester code uses two signal elements with the unit interval *T*. At the midpoint of the interval *T*/2 there is a transition from level U_H to $-U_H$, or vice versa (Fig. 2.7).

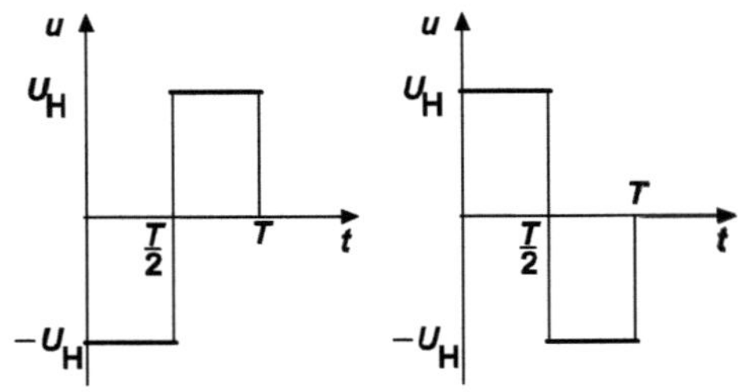

Fig. 2.7 The differential Manchester code signal components

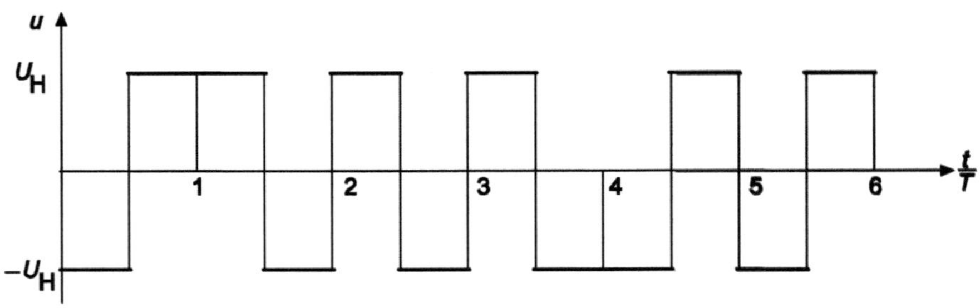

Fig. 2.8 An example of differential manchester coding

The logical state ("0" or "1") assigned to a signal element is dependent on the level of the previous signal element:

- If it is "1", the level (U_H or $-U_H$) from the previous signal element is retained
- If it is "0", the level (U_H or $-U_H$) of the previous signal element is changed (to $-U_H$ or U_H)

The line coder then converts the bit sequence 010010 into the Differential Manchester signal in Fig. 2.8.

Differential Manchester coding is an important part of data transmission in LON (see Chap. 4).

2.3 The ISO/OSI Reference Model

The ISO/OSI reference model is an international standard (ISO IS 7498) that describes how to design layered network protocols—rules that define methods of communication. It also specifies the tasks that are to be implemented at each protocol layer.

2.3.1 *Data Transmission and Communication*

Good communication technology does not automatically guarantee good communication. For example, an Englishman and German are talking about animals on the telephone. Even if they can hear each other clearly, they may not understand each other (Fig. 2.9).

The English-speaking man says "eagle", thinking, of course, of the bird of prey. But the German-speaking man hears "Igel", which is German for hedgehog but is pronounced like "eagle" in English. Clearly a translator is needed to ensure that the two parties can communicate properly.

Additionally, certain rules have to be followed, to assure an effective communication between the two men. They will not speak at the same time but instead take turns at speaking and listening. The communication will also most likely not end with one of the men just hanging up but instead, they will first exchange farewells like e.g. "Goodbye".

2.3.2 *Rules for Communicating*

Digital transmission technology on its own does not guarantee error-free communication. In addition to transmission functions, a communication system must provide a array of communication functions. A set of rules is therefore required that governs how the devices are to communicate. This set of rules is known as a protocol.

A protocol is a set of rules that governs how two devices communicate with each other.

Fig. 2.9 Problem communicating without a translator

A protocol must define, for example:

- Who is allowed to access the transmission medium and when
- How to detect data transmission errors and what happens when an error is discovered
- What happens when a receiver receives more data than it can process at a time
- How to ensure that data frames are sent to the right destination

As the whole communication process is extremely complex, there have been many attempts to try and break it down into a number of coherent, easily understandable and independent subprocesses that are connected by well-defined interfaces.

A popularly used model is the ISO Reference Model for Open Systems Inter-connection or the ISO/OSI reference model for short. ISO stands for the International Organization for Standardization.

2.3.3 The ISO/OSI Reference Model

Each OSI system typically ensures that data from an application process can be transferred to another process (Fig. 2.10).

So that application 1 in device 1 can exchange information with application 2 in device 2, both devices must have their own communication systems, consisting of hardware and software components. This system provides the applications with communication functions (services) that enable the applications to carry out their communication requests. If the communication systems of the transmitter and receiver are designed according to the ISO/OSI reference model, then they should be able to exchange data without any problems.

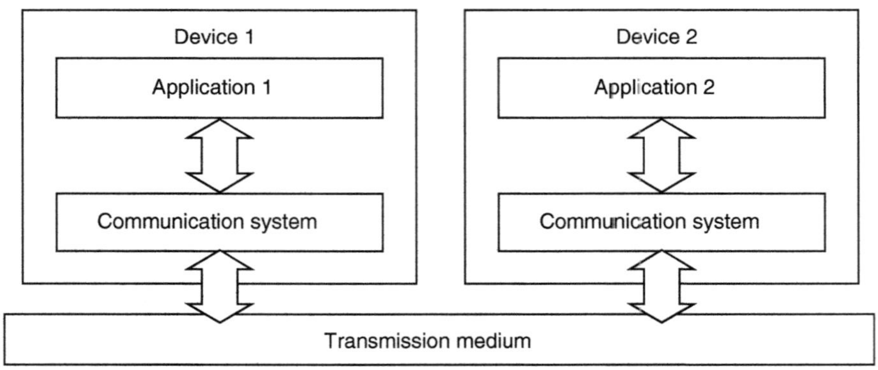

Fig. 2.10 Applications and communication systems from two communication devices

The ISO/OSI reference model classifies the communication functions into seven layers (Fig. 2.11) and is, therefore, also known as the ISO/OSI layer model.

OSI systems function as follows [3]:

- Messages travel from the source application downward from the sender through the seven layers to the transmission medium, and then upward from the transmission medium to the receiver. An OSI system is hierarchically structured and converts the data from the source device into a physical signal, and vice versa.
- Each layer provides specific communication service functions.
- Each layer provides services to the layer directly above it, for example, the physical layer (layer one) provides services to the data link layer (layer two).
- A layer has a communication interface with the layers directly above and below it.
- When a layer provides a service to the layer directly above it, it adds instructions and information to the data. For this reason, the number of bits in the data frame increases as it travels downward through each layer, and of course decreases on the way up.
- Each layer in the source OSI system communicates with its peer in the destination system. This is referred to as peer-to-peer communication. The source and destination data link layers exchange information called protocol data units (PDUs). The data is exchanged following a fixed set of rules called a protocol (e.g., the data link layer protocol). As a result, layer two in the source system is logically connected to layer two in the destination system and they communicate in accordance with their respective protocols.

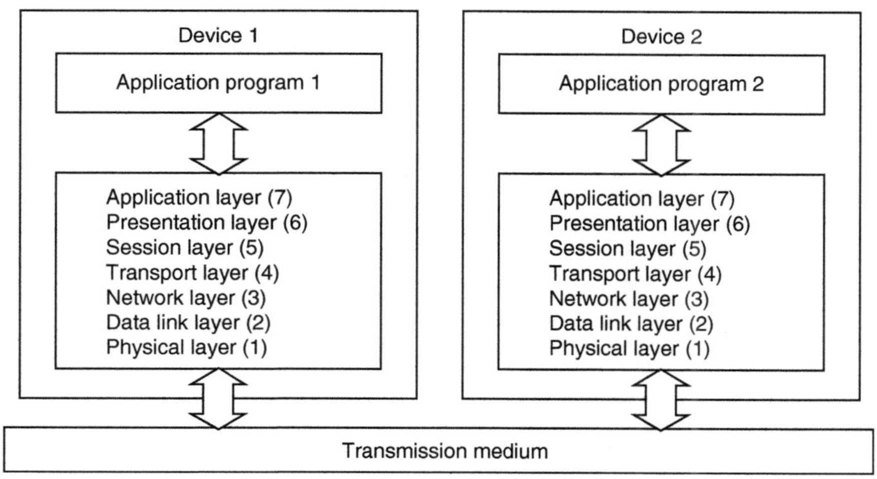

Fig. 2.11 The seven OSI layers

Each layer can be replaced as long as the functionality remains the same. For the physical layer in KNX you can use either KNX.TP (twisted pair cable), KNX.PL (data transfer over a power line) or KNX.RF (radio transmission).

A communication system does not have to implement all seven layers of the OSI model. Most field bus systems only use layers 1, 2, and 7. KNX only uses layers 1, 2, 3, 4, and 7.

2.4 Field Bus and Network Topologies

Devices (or nodes) that exchange data over a field bus or network are geometrically arranged and connected with each other in a specific way. This is known as bus or network topology. The following topologies are possible:

- Full Mesh: Each device is connected directly with all the other devices
- Partial Mesh: Only specific devices are connected with each other
- Bus: A short transmission cable connects each device to a single transmission line (backbone cable) that is terminated at both ends
- Ring: A short transmission cable connects each device to a single transmission line (backbone cable) that forms a ring
- Tree: Not only devices but also other lines are connected to a line
- Star: The devices are connected to a central station (e.g. a switch or a hub)

Additional information on topologies can be found in the chapters on KNX, LON and BACnet. Please also refer to [4] or [5].

2.5 Media Access Control Methods

In field bus and network communication all devices have direct access to the transmission channel. As a rule, all devices can receive and process the signal transferred on the channel. However, if more than one device were to send a signal simultaneously, then these signals would collide with each other. For this reason, techniques have been developed that prevent the data from colliding.

There are two ways to determine which device is allowed to transmit data on the bus or network at any one time:

- Deterministic media access
- Non-deterministic media access.

2.5.1 Deterministic Media Access

Deterministic media access control allows only one device to transmit data at any one time.

The permission to access the medium passes from one device to the next, normally in a circle. This way you can work out precisely when a device will next be allowed to transmit data (deterministic channel access).

Permission to send is allocated either by:

- One device (master) to the others (slaves), known as the master-slave method
- Passing on a token (permission) from one device to the next, known as token passing

A combination of the token-passing and master-slave methods are used in BACnet (see Chap. 5).

2.5.2 *Nondeterministic Media Access*

With non-deterministic media access, any device can send data, but it must first wait until the channel is free (idle). If it is, and a number of devices start to send data simultaneously, then a bus access conflict occurs. The device that wins access to the bus is the one with the highest priority. The loser must try and re-send its data later, but to do so it must have priority over the other devices. You can therefore not determine when this device will be able to send its data.

Examples of non-deterministic methods include:

- Carrier Sense Multiple Access (CSMA), used in LON (see Chap. 4)
- Carrier Sense Multiple Access/Collision Avoidance (CSMA/CA) used in KNX (see Chap. 3).

2.6 Exercises

Exercise 2.1
The average length of a source code word in the example of Sect. 2.2.3 is eight bits. What is the average length of a code word once it has been encoded (Shannon-Fano coding)?

Exercise 2.2
An eight-bit ASCII character is transmitted using even parity check. The receiver receives the following bit sequence: 101000100. Has a data transmission error occurred? If yes, where is the error?

Exercise 2.3
An eight-bit ASCII character is transmitted using block parity. The bits of the check character are determined using odd parity, and the parity bits of the characters are determined using even parity. The receiver receives the following sequence of bits: 101011101000101011100111011110101011.

How many ASCII characters (“real” data characters) were transmitted? Did a data transmission error occur? If yes, where?

Exercise 2.4
CRC is used to secure a sequence of bits during transmission. The check polynomial is $x^4 + x + 1$. The receiver receives the following sequence of bits: 11 0101 0011 0010. Has a transmission error occurred?

Exercise 2.5
What is the bit rate of the signal shown in Fig. 2.4 when the unit interval T of a signal element is: $T = 104$ µs?

Exercise 2.6
What is the modulation rate of the signal in Fig. 2.4 when the duration T of signal element is: $T = 104$ µs?

Exercise 2.7
What is the Manchester-coded signal for the following bit sequence? 0 1 0 0 1 0 1 0 0 0 1 1 0 1 1 0

Exercise 2.8
What is the Differential Manchester coded signal for the following bit sequence? 0 1 0 0 1 0 1 0 0 0 1 1 0 1 1 0

References

1. Gruhler G (ed) (2000) Feldbusse und Gerätekommunikationssysteme. Einführende Darstellung und detaillierter Vergleich, Steinbeis-Transferzentrum Reutlingen
2. DIN EN 60027-2: Formelzeichen für die Elektrotechnik, Teil 2: Telekommunikation und Elektronik. Nov 2007
3. Effelsberg W, Fleischmann A (1986) Das ISO-Referenzmodell für offene Systeme und seine sieben Schichten—Eine Einführung. Informatik-Spektrum 9:280–299
4. Schnell G, Wiedemann B (Hrsg.); (Hrsg.) (2006) Bussysteme in der Automatisierungs- und Prozesstechnik. Grdl., Systeme und Trends der industriellen Kommunikation. Braunschweig/ Wiesbaden: Vieweg
5. Sikora A (2003) Technische Grundlagen der Rechnerkommunikation. Fachbuchverlag Leipzig, Internet-Protokolle und Anwendungen
6. IEC 61158: Industrial communication networks - Fieldbus specifications
7. IEC 61784: Industrial communication networks - Profiles

Chapter 3
Konnex

3.1 Introduction

3.1.1 What Is Konnex?

Konnex (KNX), formerly known as the European Installation Bus (EIB), is a (industrial) building control communication system that uses information technology to connect devices such as sensors, actuators, controllers, operating terminals and monitors (see Fig. 3.1). KNX technology is designed to be used in electrical installations for implementing automated functions and processes in buildings [1].

In many KNX projects, solutions for lightning control and blinds control functions have to be implemented. A setting signal is created by a sensor as a reaction of a button push and then send to an actuator via bus in form of a data frame. The actuator then acknowledges receiving the data frame and takes an action according to the signal like e.g. switching a relay. The communication takes place between the two nodes that are necessary to realize the function. An additional communication device or DDC is not required.

3.1.2 The History of Konnex

The success story of the European Installation Bus (EIB) began in May 1990 when the European Installation Bus Association (EIBA) was founded in Belgium by 15 European building control technology manufacturers. The goal of EIBA was to

---

The importance of the European Installation Bus KNX has already been pointed out in Sect. 1.7.2.1. This chapter will provide you with a detailed description of KNX.

H. Merz et al., *Building Automation*, Signals and Communication Technology,
https://doi.org/10.1007/978-3-319-73223-7_3

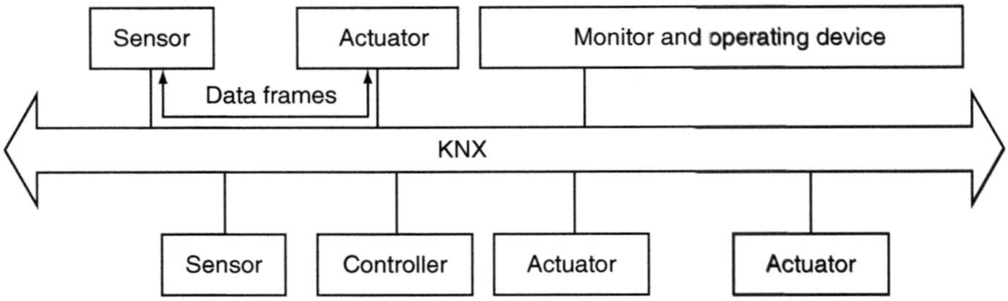

Fig. 3.1 Building control devices connected over Konnex

Fig. 3.2 KNX Logo

advertize and promote the use of EIB as an international standard. In May 1999, the fusion of EIBA, BatiBUS Club International (BCI) from France and the European Home Systems Association (EHSA) from Switzerland formed the new Konnex Association (www.konnex.org). Ever since, EIB is marketed under the name KNX. The KNX logo can be seen in Fig. 3.2. The Konnex Association is a group of more than 100 building control technology companies with a market share in Europe of over 80%. The office of the German KNX group (www.knx.de) is located in Frankfurt as part of the German Electrical and Electronic Manufacturers' Association (ZVEI).

In the following years a vast number of KNX standards was created:

- European Standard CENELEC EN 50090
- European Standard CEN EN 13321-1 and 13321-2
- International Standard ISO/IEC 14543-3
- Chinese Standard GB/Z 20965

The connection to other building automation systems (via BACnet) is standardized as well:

- US ANSI/ASHRAE Standard 135
- EN ISO 16484-5

KNX is the only open standard for building control technology worldwide. In this case "open" means, that devices from different manufacturers can communicate with eachother.

3.1.3 *The Benefits of Konnex*

3.1.3.1 Functions in Buildings and Communication Participants

To implement various building functions [2, 3] such as

- Input and output functions
- Processing functions
- Management functions
- Operational functions

the different participants

- Sensors (e.g. brightness sensors, presence motion detectors, switch sensors)
- Actuators (e.g. relays, dimmers, motors)
- Control and regulation devices (e.g. thermostats)
- Operating and monitoring (visualization) devices such as control panels

need to be able to communicate with eachother [4–6].

3.1.3.2 Conventional Building Control Technology

In conventional building installations, each system (lighting, heating, air conditioning, etc.) is planned and implemented by a specialist company. The sensors and actuators used are usually connected to control and display panels via point-to-point connections. This leads to a considerable increase in:

- Planning effort
- Necessary wiring and respective installation
- Troubleshooting for commissioning and maintenance

Furthermore, the more wires used the greater the fire risk. Connecting these systems so that they can all be centrally operated would require a vast amount of wiring. Increasing the functionality and usability of a conventional building installation makes it more complex, complicated and expensive. KNX provides a way of resolving these issues.

3.1.3.3 Building Control Using Konnex

Nowadays there are devices with standardized KNX connection for every system. This simplifies the planning and implementation of building functions and provides superior functionality, flexibility and comfort without any additional effort and expense. By setting the appropriate parameters—which can be modified at any time—you can tell the device exactly what it is supposed to do. This makes a KNX system extremely flexible, allowing you to adjust and expand it to meet new requirements at any time.

Unfortunately, KNX products cost far more than those used in conventional installations. Generally, the investment is only worthwhile if several systems are to be connected with each other or if an installation needs to be flexible enough so that it can be quickly and effectively modified to meet any future changes in use or requirements.

3.1.4 Reasons for Learning About KNX

The important role KNX plays in building automation not only in Europe but worldwide is reason enough to learn more about it. But here are a few more:

- KNX provides the ideal introduction to the world of bus technology. Many important features and terms used in current bus systems are also used in KNX such as bit rate, media access control, data frames, protocols, etc.
- Many of the problems that can be solved by KNX occur in our everyday lives. This makes it easy to become familiar with how to solve automation-related problems using KNX.
- Many professional training schools teach bus systems, particularly KNX, as part of their syllabus.
- KNX is simple and straightforward to plan, program and implement. Approximately ten percent of companies in the electrical and electronics industry can plan and install KNX systems. Many engineering companies take on KNX projects, which mean that their energy and building system engineers need to have a good working knowledge of KNX.
- Using IP Gateways, KNX systems are increasingly being integrated into the network infrastructure of buildings (TCP/IP networks), allowing them to be accessed externally via the Internet. KNX also enables you to connect electrical installations to the Internet. In this respect, learning about KNX provides a sound in-depth introduction to modern automation technology.

3.2 Conventional Installation Technology

This section explains how a simple lighting control system can be implemented using conventional electrical installation technology (on/off, changeover, and crossover switches). You will see that the system becomes rapidly more complex if you want to be able to turn a light on or off from more than one place. You cannot implement more complex functions such as time-dependent applications with the basic circuitry used in conventional installations. Instead, you need special devices such as stairwell timer switches or latching relays. In such cases, it is always worth considering using KNX.

3.2.1 Safety Instructions

You can buy circuit components such as on/off, changeover and crossover switches in any hardware store. As a result, you might think about installing and testing a circuit yourself. Before you do so, however, please read the following safety information [7]:

Only qualified electricians are permitted to work on an electrical installation.

- The person in charge of working on an electrical installation must ensure that all applicable regulations are complied with and is therefore liable for any damages. Insurance cannot be claimed for damages resulting from a faulty electrical installation.
- Never work on live devices or systems! Always disconnect the power before you start working, either by pulling the plug from the socket or turning off the electricity from the fuse box!
- Put up a warning sign saying that you have turned off the electricity to make sure that no one else switches it back on before you have finished!
- Always make sure the electricity is turned off before you start working with a voltmeter!

3.2.2 Task: Stairwell and Corridor Lighting in an Apartment Building

The following example highlights the differences between conventional installation technology and building control with KNX.

A real-estate developer wants to build a three-story apartment building. Each floor will have a corridor with two doors to each apartment. A light switch for the corridor lighting will be fitted next to each door (see Fig. 3.3). A stairwell will provide access to each floor. The stairwell will have a light switch on each floor (next to the door) to turn the three stairwell lights on or off simultaneously.

A typical scenario: A first-floor resident leaves his or her apartment through the door on the right, turns on the corridor light using the light switch on the right, walks to the stairwell, turns off the corridor light with the light switch on the left, walks down the stairs, and then turns off the stairwell lights using the stairwell light switch on the first floor.

The real-estate developer asks an electrician for advice on how to implement the lighting system. The electrician comes up with the following proposals:

- Conventional lighting control using changeover switches and crossover switches
- Lighting control using KNX

Section 3.2 will explain how to implement lighting control using conventional installation technology.

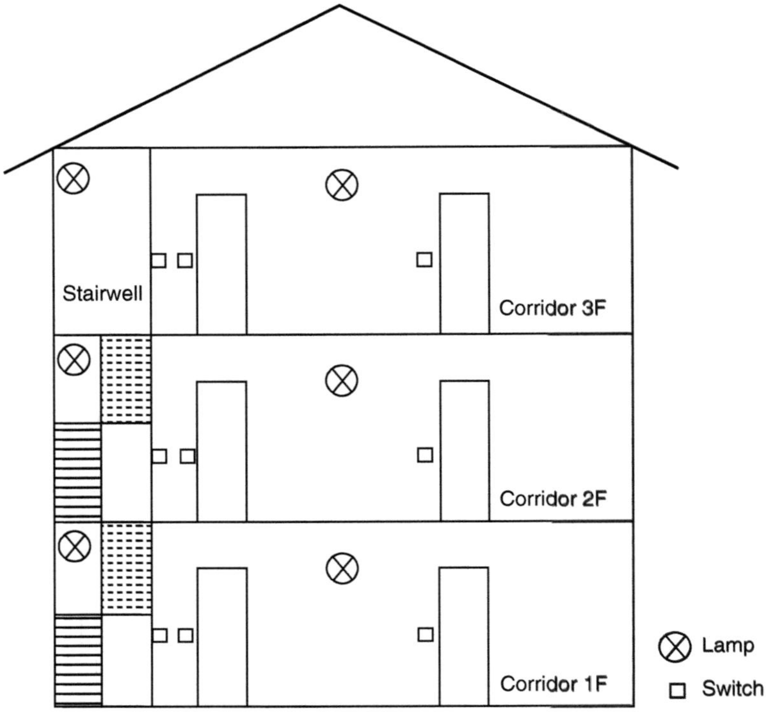

Fig. 3.3 An apartment building with stairwell and corridor lighting

Section 3.10 will show you how this lighting system can easily be implemented using KNX. Konnex even enables you—without any additional effort or expense—to implement functions, such as a central on/off switch, that are far more complex than the switching function requested by the real-estate developer, and which provide far more benefits than conventional circuitry.

3.2.3 On/Off Switching Circuits

If you want to be able to turn a light on or off from one point, then you need an on/off switching circuit (Fig. 3.4).

An on/off switching circuit consists of one unipolar or one bipolar on/off switch. (The bipolar on/off switch is far safer because it completely separates the load from the power line.)

The connection point L of the unipolar on/off switch is connected to the L1 wire of the low-voltage power line (1–60 Hz, 120 V/1–50 Hz, 230 V). Connection point 1 or 2 is connected to the lamp. The lamp is then connected to the neutral wire (N)

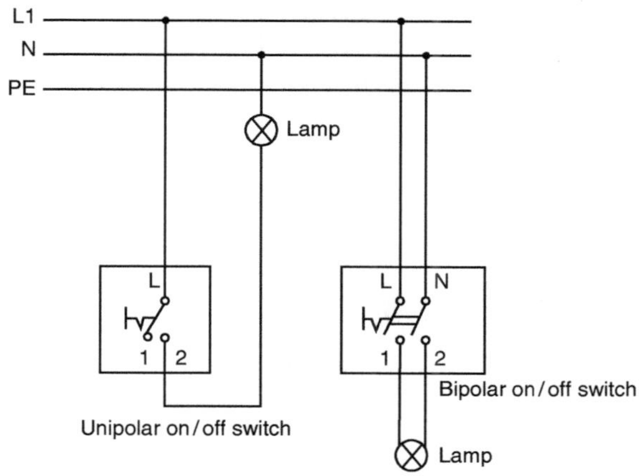

Fig. 3.4 Unipolar and bipolar on/off switches

Table 3.1 On/off switch functions as shown in Fig. 3.7

	On/off switch	Lamp
Switch setting	Left (1)	Off
	Right (2)	On

(Fig. 3.4 left). The protective earth (PE) conductor can also be connected to the lamp, if necessary. The bipolar on/off switch is set up in much the same way (Fig. 3.4 right).

Table 3.1 shows the functions of a unipolar on/off switch as shown in Fig. 3.4. The bipolar on/off switch works in the same way.

You can see from the configuration and functionality of an on/off switch circuit how to implement on/off switching using conventional electrical installation technology—namely by opening or closing switches that are on the live wire. As a result, the flow of information and flow of energy are not separate.

To turn a lamp on or off from one point, you just need to connect a unipolar on/off switch to the live wire. You do not need any additional wires. However, this is not the case with changeover and crossover switching circuits, as the following sections will explain.

3.2.4 *Changeover Switching Circuits*

If you want to able to turn a lamp on or off from two points, then you need a changeover switching circuit (see Fig. 3.5) (Table 3.2).

As with an on/off switching circuit, the on/off switching in a changeover switching circuit is implemented by opening or closing switches connected to the live wires. Here, however, you need more than just two changeover switches. You also need two additional "information" wires to connect the two changeover switches. These wires are also used to transfer electricity at the same time. As a result, the flow of information and the flow of energy are not separate.

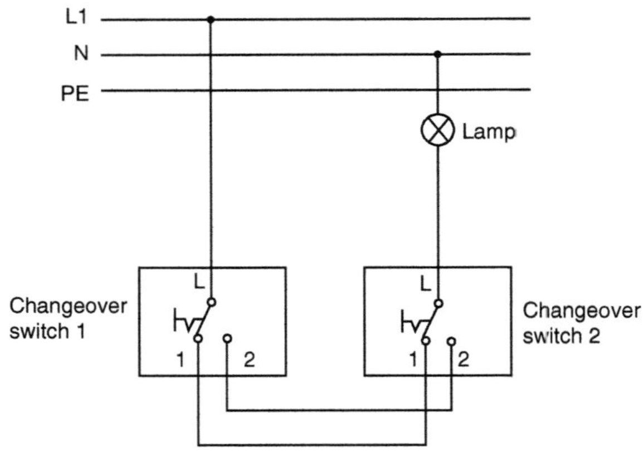

Fig. 3.5 A changeover switching circuit

Table 3.2 Changeover switching circuit functions, as shown in Fig. 3.5

	Changeover switch 1	Changeover switch 2	Lamp
Switch setting	Left (1)	Left (1)	On
	Left (1)	Right (2)	Off
	Right (2)	Right (2)	On
	Right (2)	Left (1)	Off

3.2.5 *Crossover Switching Circuits*

To turn a lamp on or off from three points, you need a crossover switching circuit, as shown in Fig. 3.6.

A crossover switching circuit consists of two changeover switches and one crossover switch. To convert a changeover switching circuit to a crossover switching circuit, disconnect the two wires connecting the two changeover switches and install a crossover switch.

Make sure that the crossover switch's two internal switches S1 and S2 are connected to each other. There are only two possible settings:

(a) S1 left (1)/S2 left (2)
(b) S1 right (2)/S2 right (1)

The following connections are required for the switching circuit to work:

- The connection point L of changeover switch 1 must be connected to the L1 wire of the low-voltage power line (1–60 Hz, 120 V/1–50 Hz, 230 V).
- The connection points 1 and 2 of changeover switch 1 must be connected to the crossover switch's connection points L and L (or 1 and 2). The remaining connection points on the crossover switch must then be connected to the connection points 1 and 2 on changeover switch 2.

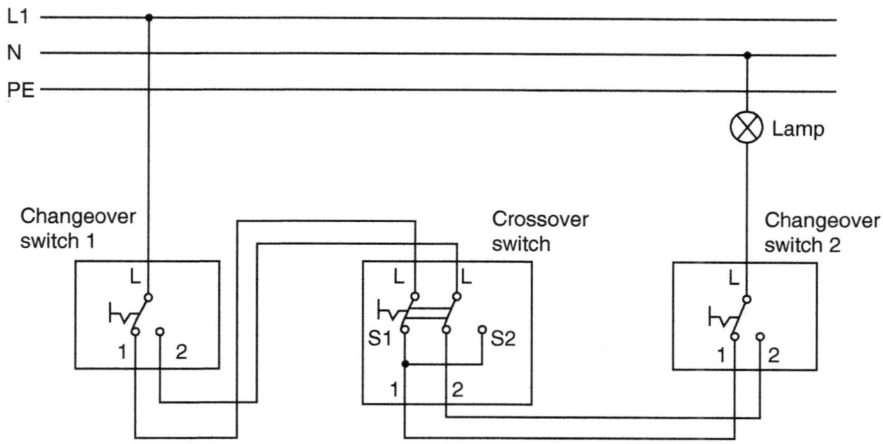

Fig. 3.6 A crossover switching circuit

- The connection point L of changeover switch 2 must be connected to the lamp, and the lamp's second connection point must be connected to the neutral wire (N).
- Connect the PE conductor to the lamp if necessary.

Depending on the settings of the changeover and crossover switches, the voltage on the power line is applied to the lamp, which is then lit by the electrical current. Table 3.3 shows the functions of the crossover switch shown in Fig. 3.6. When the lamp is off, any of the three switches turns it on; when the lamp is on, any of the three switches turns it off.

The configuration and functionality of a crossover switching circuit also shows that the switching information is implemented by opening and closing switches that are connected to the live wire. As with the changeover switch example, additional wires are needed to connect the changeover switches to the crossover switch. Again the information and energy flows are not separate.

Table 3.3 Crossover switching circuit functions, as shown in Fig. 3.6

	Changeover switch 1	Crossover switch	Changeover switch 2	Lamp
Switch setting	Left (1)	Left	Left (1)	On
	Right (2)	Left	Left (1)	Off
	Left (1)	Right	Left (1)	Off
	Left (1)	Left	Right (2)	Off
	Right (2)	Right	Left (1)	On
	Right (2)	Left	Right (2)	On
	Left (1)	Right	Right (2)	On
	Right (2)	Right	Right (2)	Off

3.3 Overview of Konnex

This section will start by providing an overview of the aspects of KNX that are dealt with in this chapter (Fig. 3.7).

1. When setting up a KNX system you must first decide which transmission medium to use (Sect. 3.4). Twisted pair cable (KNX.TP) is the most common.
2. Then select the bus devices needed for all the required functions (Sect. 3.5). It helps if you also have an understanding of the devices' "external" and "internal" hardware (see Sects. 3.12.1 and 3.12.2).
3. The KNX devices need to be assigned unique physical addresses, defining the topology of the system as KNX devices are assigned to lines and areas. Additionally, installation guidelines (Sect. 3.6.5) need to be considered e.g. the maximum length for a KNX.TP line (1000 m).
4. Finally, you configure and program the KNX bus devices using the ETS 5 program (Engineering Tool Software Version 5, see Sect. 3.13.6). Here an important step is the assignment of communication objects to group addresses, defining the communication relation between sensors and actuators (Sect. 3.8).

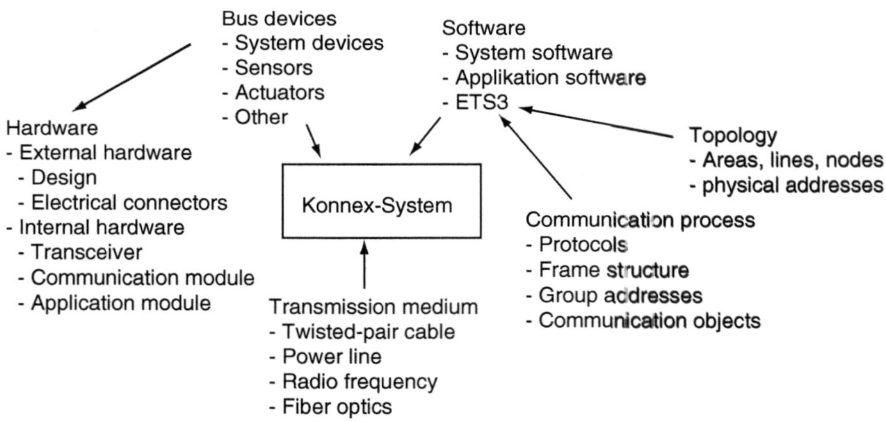

Fig. 3.7 Overview of KNX

3.4 Transmission Media and Features of KNX.TP

3.4.1 Transmission Media

Different types of transmission media can be used to transfer data from one device to another [8]:

- Twisted pair (KNX.TP)
- Power line (KNX.PL)
- Radio frequency (KNX.RF)
- Ethernet (KNXnet/IP)

The data transmission over these media is standardized [8]. Different manufacturers also offer transmission via infrared (KNX.IR) or optical fiber [9]. When a KNX device wants to send data, the bit information to be transferred is converted in a physical signal like e.g. voltage signal, radio frequency signal or light signal using the respective transceiver (see Sect. 3.12.2.2). The receiver converts the physical signal back to a bit sequence.

3.4.2 Criteria for the Choice of a Transmission Medium

The most frequently used medium especially in new buildings is transmission via twisted pair (KNX.TP) as it is the cheapest alternative if an additional data cable can be installed without any problems. However, in many cases an electricity grid is present but no additional cable should be installed for various reasons, e.g. in churches or landmarked buildings. In this case, power line (KNX.PL) is used. For KNX.PL, the data signals are added to the sinusoidal voltage of the energy supply network [10]. For KNX.RF and for the transmission via infrared no additional bus lines have to be installed either, since radio frequency signals and infrared signals are used for transmission. KNXnet/IP is used to connect KNX devices with TCP/IP networks of building automation, e.g. for the communication with control and monitoring stations. Optical fiber is used to cover large distances when connecting multiple buildings. A major benefit of optical fiber is that no lightning protection or overvoltage protection has to be installed.

3.4.3 Characteristics of KNX.TP

Important aspects of the most common KNX variant KNX.TP, especially those regarding bus lines and bus signals, are described in the following section.

3.4.3.1 Bus Lines

The Konnex Association has certified multiple KNX lines [8]. The classic KNX line (Fig. 3.8) is referred to as YCYM 2 × 2 × 0.8 and is part of the group of control lines.

It contains four copper wires with a diameter of 0.8 mm (2 × 2 × 0.8) insulated with PVC (Y), twisted with eachother and shielded with a film (C). The two additional letters (YM) refer to the type of line. It is a non-metallic sheathed cable (M) with a (green) protective shell made from PVC as well (Y).

The red (+) and black (−) pair of wires is used for supplying the devices with power and, at the same time, transmitting data. The yellow and white pair of wires is used as a reserve, e.g. for an additional bus line or to supply the devices with additional energy if required.

Twisted-pair cables can be flush or surface-mounted in dry, moist and wet rooms. The same installation specifications that apply to power lines also apply to twisted-pair cables [7].

3.4.3.2 Bus Signal

During the transmission via KNX.TP a KNX data frame is converted to a voltage signal by the transmitter and converted back to a bit sequence by the receiver. In order to do so, a special component like e.g. a TP-UART-IC (see Sect. 3.12.2) is necessary. The signal is a differential voltage between the red positive wire (called A line) and the black negative wire (called B line). The stress flow over time can be measured using e.g. an oscilloscope. As this process should be potential-free, the oscilloscope has to be connected to the low voltage grid via the isolating transformer.

Fig. 3.8 Classic KNX.TP line YCYM 2 × 2 × 0.8 with the Konnex Association's certification number 1/39/92

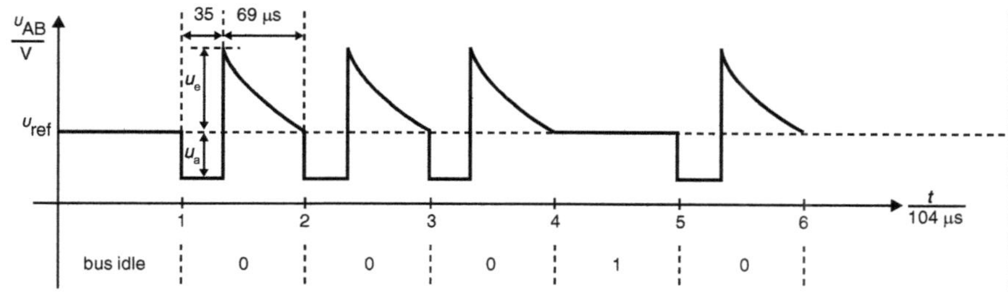

Fig. 3.9 Stress flow for KNX.TP

Figure 3.9 shows the theoretical flow of a KNX.TP signal using the example of the first five bits from a data frame signal. The meaning of zero-bits and one-bits is explained in Sect. 3.10.

The difference in potential $\varphi_A - \varphi_B = u_{AB}$ between A line and B line is processed by the receiver. When there is no bus activity present (bus idle) or when a one-bit is sent, the differential voltage during the bit period $T_{bit} = 104\ \mu s$ has a constant value u_{ref} between 32 and 21 V. When a zero-bit is sent, the change in voltage is correlated to this reference DC voltage.

In case of a zero-bit, the transceiver of the transmitter actively lowers the voltage level during the time period of 35 µs. For the falling edge and the constant level starting at point t_A, the following applies:

$$(u_{ref} - 10.5\,V) \leq u_a \leq (u_{ref} - 0.7\,V)$$

For the constant level and the rising edge until point $t_{B,}$ the following applies:

$$(u_{ref} - 10.5\ V) \leq u_a \leq (u_{ref} - 0.5\ V)$$

After that, there is an equalizing pulse releasing the energy mainly stored in the inductor and the following applies:

$$u_{ref} \leq u_e \leq (u_{ref} + 13\,V)$$

At the end of a bit period of 104 µs, the following has to apply to the zero-bit signal element:

$$(u_{ref} - 0.35\,V) \leq u_{AB} \leq (u_{ref} - 1.8\,V)$$

If a zero-bit is sent prior to a one-bit, a linear voltage drop can occur at the one-bit signal element during the bit period $T_{bit.}$ In order to safely identify the one-bit, the following conditions have to be met:

- The voltage u_{AB} can not fall with a rate higher than 400 mV/ms
- At the end of the bit period the following has to apply: $(u_{ref} - 2\,V) \leq u_{AB} \leq (u_{ref} + 0.3\,V)$

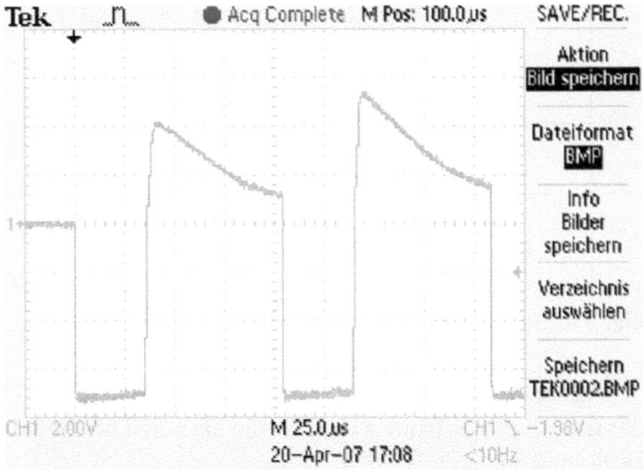

Fig. 3.10 KNX.TP stress flow (measured)

The stress flow depicted in Fig. 3.9 is an idealization! Depending on the number of communication participants, the distance to the power supply (drop of voltage lengthways of the line, line capacity) and external interfering voltages, the actual result can differ drastically (Fig. 3.10) from the ideal form.

The receiver components can still progress deformed signals as long as they are within the defined tolerance range. By using couplers and line repeaters (Sect. 3.6.4) the signals are regenerated when they are transferred to another line or line segment and can therefore be safely detected even if the receiver is far away from the transmitter.

3.4.3.3 Bitrate

Based on the bit period T_{bit} = 104 µs that is necessary to transmit one bit, the bitrate r_{bit} for KNX.TP can be calculated as follows:

$$r_{\text{bit}} = \frac{1}{T} = \frac{1\,\text{bit}}{104\,\mu\text{s}} \approx 9615\frac{\text{bit}}{\text{s}} \approx 9.6\frac{\text{Kbit}}{\text{s}}$$

3.5 KNX Bus Devices

In order to install a KNX system, different kinds of bus devices are required. To get an overview of the commonly used KNX devices, you can refer to the manufacturers' catalogs or websites. Especially the websites contain a lot of technical information as well as application examples. If all manufacturers are considered, the number of possible devices can easily reach a few thousand. It is necessary to be very familiar with the desired function and location in the system to find the ideal products.

3.5.1 Types of Bus Devices

The KNX products listed in the manufacturers' catalogs are usually sorted by transmission medium e.g. KNX.TP or KNX.PL. They can also be classified by design or the location where they are installed:

- Flush-mounted (FM): suitable for flush-mounted fittings (in standard flush-mounted sockets) such as the bus coupler shown in Fig. 3.14.
- Surface-mounted (SM): suitable for surface-mounted fittings.
- Built-in (BI): suitable for electrical conduits, hollow floors and suspended ceilings.
- Rail-mounted (RM): suitable for installation in control cabinets by clipping them on to a DIN rail (see Fig. 3.12).

Eventually, the devices are categorized in the following groups:

- System components (power supply units, accumulators, line couplers, backbone
- couplers, line repeaters, bus couplers, and RS-232 or USB interfaces)
- Sensors (switch sensors, presence detectors, glass break detectors)
- Actuators (switch, blind or shutter actuators)
- Visualization (LC displays, control panels)
- System accessories (terminals)
- Accessories (flush-mounted sockets, labeling tools).

KNX devices can also be modular or compact. Modular devices consist of an application module plugged on to a bus coupler (see Fig. 3.11).

With compact devices the application module and bus coupler are housed in the same unit (see Fig. 3.13).

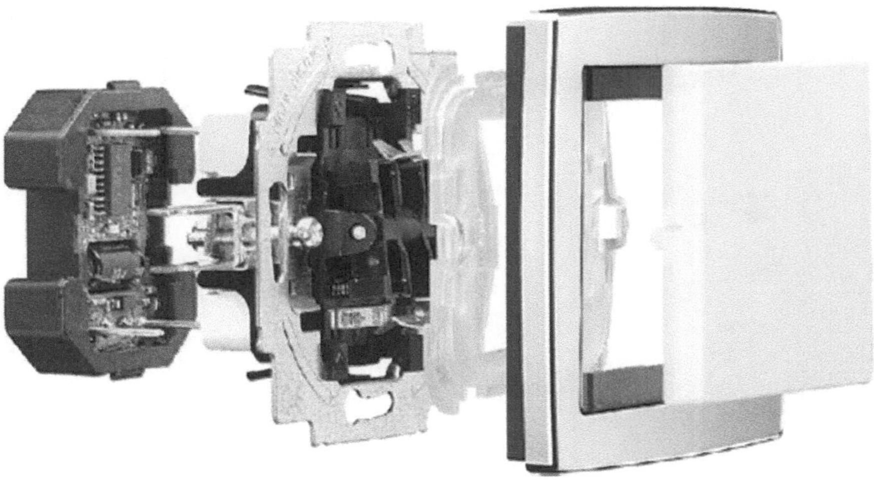

Fig. 3.11 A modular KNX.PL device [Busch-Jaeger Elektro] on the far left: bus coupler

3.5.2 *Frequently Used Bus Devices*

Below are three devices that are often used in KNX systems, particularly for lighting control, and which represent the variety of KNX products currently available:

- Power supply (PS) unit with integrated choke
- Six-gang switch actuator
- Four-gang switch sensor.

3.5.2.1 Power Supply Units with an Integrated Choke

The power supply with an integrated choke (Fig. 3.12) is connected to the low voltage power line and supplies the bus with a direct voltage of usually 30 V DC. This does not mean however, that all devices connected to the bus have an input voltage of 30 V—the further away the device is from the power supply, the lower the input voltage down to the minimum voltage required for a bus device of 21 V DC. In order to guarantee this voltage, the maximum length of line segments has to be kept in mind (see Sect. 3.6.5). The power supply has an additional output for

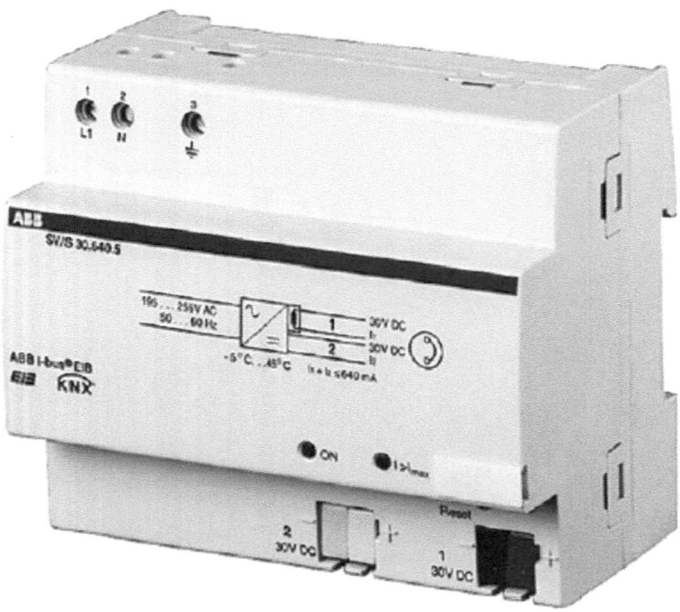

Fig. 3.12 A power supply unit 640 mA (RM) with an integrated choke and two 30 V DC outputs [9]

supplying a second line with 30 V DC. However if this line should be used for data communication as well, another separate choke is required.

3.5.2.2 Switch Actuators (Six-Gang)

A six-gang switch actuator (Fig. 3.13) is a rail-mounted device that has six potential-free output channels A–F, which can each switch one electrical load (230 V AC/10 A or 400 V AC/6 A).

Each output functions as an on/off switch. Each on/off switch can be connected to the L1 wire of the low voltage network and to the load using screw terminals (1–2 to 11–12) (see Figs. 3.13 and 3.30).

3.5.2.3 Switch Sensors (Four-Gang)

A four-gang switch sensor (Fig. 3.14, right) is an application module that is fitted (Fig. 3.15) on to a bus coupler (Fig. 3.14, left).

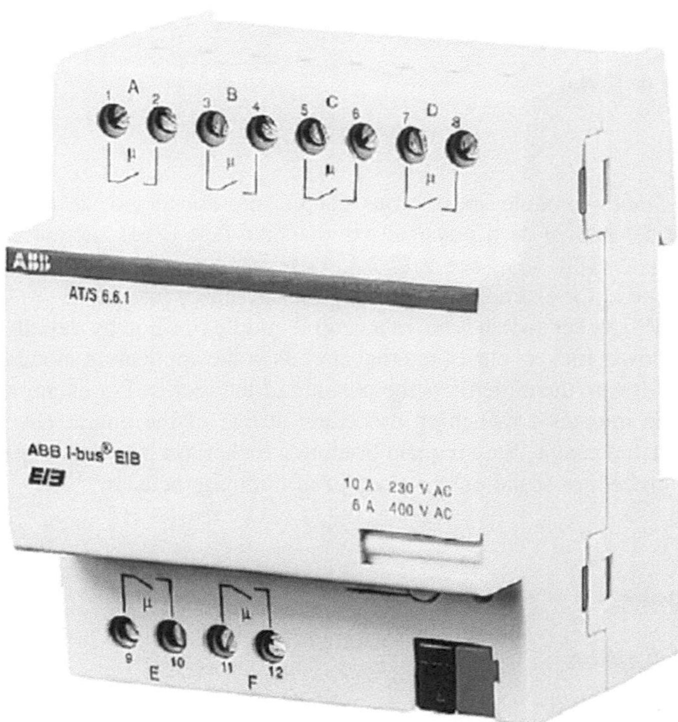

Fig. 3.13 A six-gang switch actuator (RM) [9]

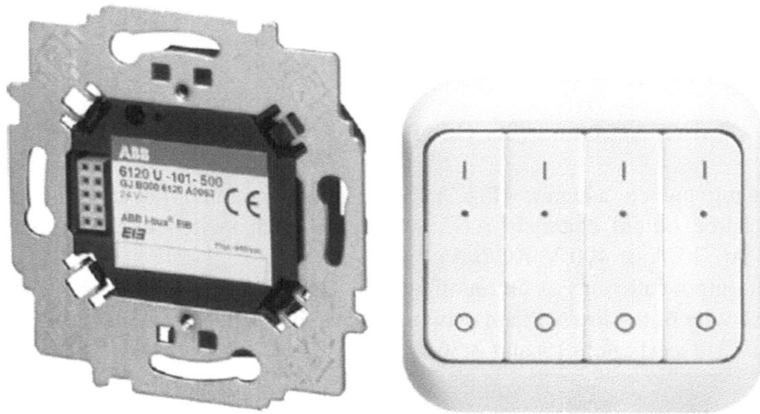

Fig. 3.14 A bus coupler (BC) and four-gang switch sensor [9]

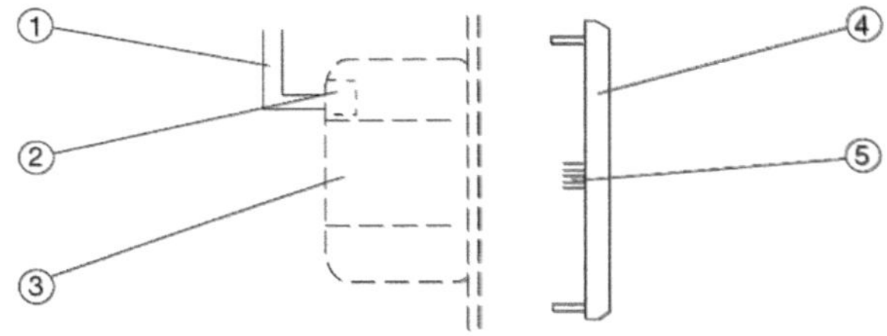

Fig. 3.15 Connection of a bus coupler (FM) and four-gang switch sensor [9]. 1 bus line, 2 bus terminal, 3 bus coupler (FM), 4 application module, 5 connector (10- or 12-pin)

The application module and the bus coupler are connected via a 10-pin connector (2×5), known as a physical external interface (PEI). 12-pin connectors (2×6) exist as well, see Sect. 3.12.2.4. Depending on your requirements, you could also attach, for example, a two-gang rocker sensor instead.

The sensor's rocker switches are in a neutral middle position by default. Pressing the upper or lower rockers closes internal circuits in the application module. Special electronics/software then interprets the position of the rocker. For example, tapping a rocker once initiates a switching procedure at one of the output channels of a six-gang switch actuator. Pressing and holding a rocker, on the other hand, initiates a dimming procedure at the output channel of dimming actuator.

3.6 Topology

3.6.1 Definition

Like a conventional electrical installation, a KNX.TP installation needs to have a power line network to provide the loads (lights, engines, but also KNX devices such

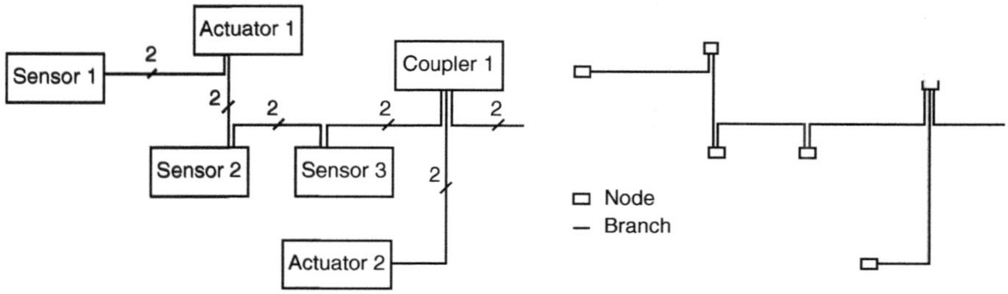

Fig. 3.16 Example of topology. A KNX system (left) and its network (right)

as power supply or touch panels) with electricity. But in addition to that, it has a communication network (the KNX) that supplies bus devices with electricity but also transmits data. Both networks are galvanically isolated, shown by the separate cables. As a result, a KNX.TP installation includes planning and installing not only a power line network but also a communication network.

To ensure that a KNX system fulfills the customer's requirements, the designer first calculates how many system components, sensors, actuators, etc. are needed. Then he or she determines where the KNX devices are to be installed in the building, and how they are going to be connected with each other via a bus cable to a communication system (see Fig. 3.16, left). The structure of this system is called *topology*.

Topology describes the structure of a system in terms of how the devices (nodes) are connected with each other. It is illustrated using network diagrams.

A network diagram (Fig. 3.16, right) consists of nodes and branches. The nodes in the communication network represent the KNX devices, which are directly connected with at least one other KNX device. A branch as a connection of two nodes is a (twisted pair) bus line of a certain length as seen in Fig. 3.16 to the left. For KNX, these branches are called lines or line segments.

3.6.2 Nodes, Lines, and Areas

A KNX system is hierarchically structured. The topology is based on the structure of conventional building installations—known as tree topology (see Fig. 3.17).

- Nodes (N) are assigned to a line (L)
- Several lines are connected to a main line (ML) via couplers (C) and form an area (A)
- Several areas are connected with each other by connecting the main lines to the backbone line (BL) via couplers.

An area represents, for example, the first floor of a building. On each first floor corridor are the lines to which the bus devices in the neighboring rooms are

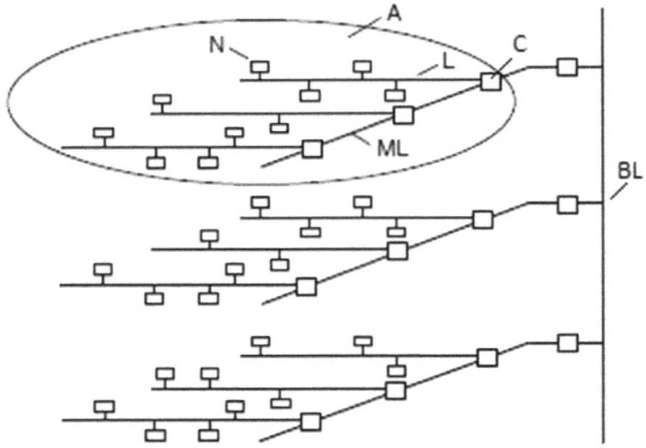

Fig. 3.17 KNX tree topology

connected. Smaller systems often only consist of a few nodes that can be assigned to one single area and to one single line. A device must be assigned to a line and an area! The number of nodes, lines and areas depends, for example, on the size of the system and the number of bus devices that are to be installed.

Sometimes more lines than actually necessary (as per size of the building or number of devices) are used. This is the case if, for example, a redundant line is installed in order to increase availability of functions or if a line should be protected from data frame traffic caused by connected devices by including a regulation device. In this case, the regulation device needs an idle bus in order to transmit the regulation information on time.

In the KNX.TP-version TP1-64, a line consists of up to four line segments, each able to include 64 bus devices. That means that a TP1-64 line can handle 256 devices. In the KNX.TP-version TP1-256, a line consists of only one line segment that can handle 256 devices. More devices aren't possible, since only 8 bits are available for addressing.

When planning a KNX system, it is convenient to keep an expansion reserve of 20% so that no lines have to be added should the system expand in the future. A line with one line segment in version TP1-64 should consist of no more than 50 devices. For lines with multiple line segments, this applies to every line segment respectively.

Lines or line segments and areas are connected using line repeaters, line couplers and bus couplers. The coupler is the same device in any case, its designation is dependant on its position in the topology and on its application program (see Sect. 3.6.4). Sensors, actuators and couplers are the actual participants of the data communication in a KNX system. They are assigned a physical address so that they can be distinctly identified in the system. Physical addresses are explained in Sect. 3.7. For electricity and communication, every line (line segments, backbone lines, main lines) requires its own power supply unit (with a choke).

3.6.3 *Power Supply Units*

A power supply (PS) unit supplies the nodes on a line with electricity at a nominal voltage of 24 V DC and, together with the choke, enables the nodes to communicate with each other on the line. There are two types of power supply units: a PS that supplies a maximum current of 640 mA for up to 64 nodes (see Fig. 3.12), and a PS with a maximum current of 320 mA for up to 32 nodes. The PS you choose depends on how many nodes there are, or you plan to have, on each line.

One advantage of using one PS per line is that if a PS fails, then only the nodes on that particular line are affected—they will no longer be able to communicate—whereas the nodes on the other lines will be unaffected. Of course, if there is a problem on a main line or the backbone line, a vast amount of communication between the lines and areas will be disrupted.

Line repeaters (LR), Line couplers (LC) and backbone couplers (BC) are always powered by the PS on the subordinate line, which means (see Fig. 3.19):

- LRs are powered by the PS on line segment 1, 2 or 3
- LC by the PS on line segment 0
- BC by the PS on the main line

For electrical reasons (in version TP1-64), the maximum number of nodes per line, main line and backbone line is as follows:

- 64 sensors/actuators or 63 sensors/actuators plus one LC in a line (with one line segment 0)
- 63 sensors/actuators plus one LR in the line segments 1–3
- 64 sensors/actuators or 63 sensors/actuators plus one BC in a main line
- 64 sensors/actuators in a backbone line

Coupling line segments or lines only works correctly, when both the subordinate and the superior line include a PS!

Example: Sensor 1.1.1 is connected to actuator 1.2.10 via LC 1.10 and LC 1.2.0. For the system to work, three PS are required: One PS for line 1.1, one PS for main line 1.0 and one PS for line 1.2.

3.6.4 *Couplers*

A coupler (Fig. 3.18) is a system component and can be used as a line repeater (LR), a line coupler (LC) or a backbone coupler (BC). Its role is defined during configuration by assigning a specific physical address (see Sect. 3.7) and selecting the corresponding application program (coupler or repeater).

The lines and line segments that are connected with each other by LCs, BCs and LRs are electrically isolated. This galvanic isolation is maintained even when LCs, BCs and LRs are installed—only a data connection is established.

Fig. 3.18 Coupler [9]

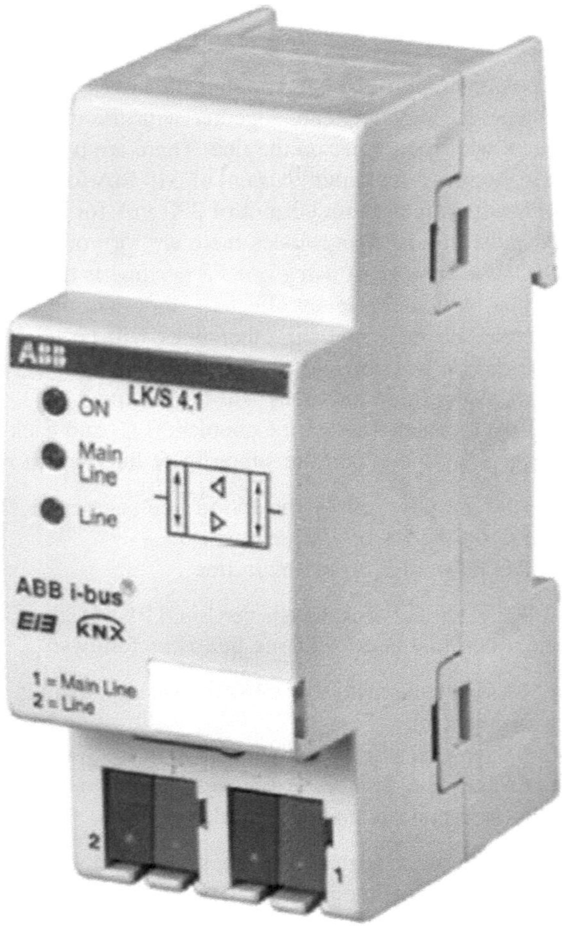

3.6.4.1 Line Repeaters

You can use a coupler as a line repeater to expand a line that only has one line segment (LS, with the number 0) to include more than 64 devices or to increase the length of the line. You can add a maximum of three line segments (numbers 1–3) (see Fig. 3.19).

A line can have a maximum of 256 nodes, comprising four line segments (LS 0–LS 3) with 64 nodes on each segment. The LSs 1, 2, and 3 must be connected in parallel. The reason for this is that after a normal data frame has passed through six couplers (LC, BC or LR) it is no longer forwarded (see Sect. 3.10.5).

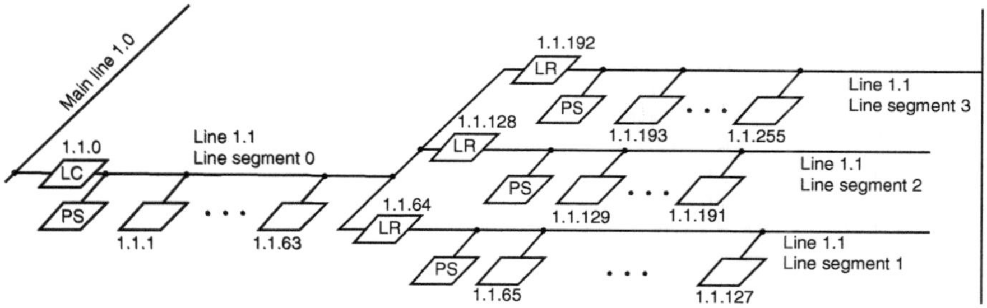

Fig. 3.19 A line with three line repeaters and four segments

For a line's maximum configuration the following rules apply:

- Line segment 0 can have a maximum of 64 actuators/sensors, or 63 sensors/actuators and one LC
- Line segments 1–3 can each have a LR and a maximum of 63 actuators/sensors

In the maximum configuration, the LRs are assigned the physical addresses: 1.1.64, 1.1.128 and 1.1.192. However, if only 50 sensors/actuators are installed per line segment, other addresses are available, for example, 1.1.51, 1.1.101, 1.1.151.

A line at its maximum configuration has 253 sensors/actuators (or 252 if another LC is connecting it to the main line).

For KNX.TP version TP1-256, there is only one line segment with a maximum of 256 sensors/actuators or 255 plus one LC. In this case, couplers as line repeaters are not required.

3.6.4.2 Line Couplers

Line couplers (LC) connect the lines within an area (Fig. 3.20) to the area's main line. In order to make a coupler a line coupler, it is assigned a special physical address (1.1.0, 1.2.0 etc., see Sect. 3.7.1.1) and the application program coupling.

3.6.4.3 Backbone Couplers

Backbone couplers (BC) connect the main lines to the backbone line (Fig. 3.20). In order to make a coupler a backbone coupler, it is assigned a special physical address (1.0.0, 2.0.0 etc., see Sect. 3.7.1.1) and the application program coupling.

3.6.5 Installation Guidelines

You must follow the installation guidelines when arranging the KNX devices in a building to, firstly, ensure that the devices receive the required minimum voltage

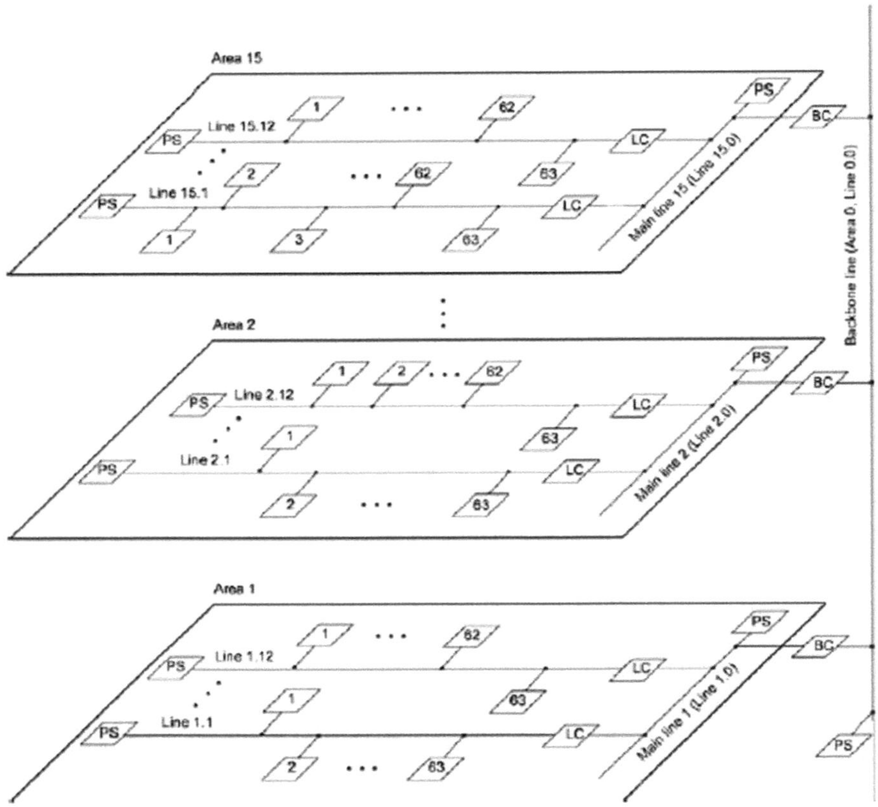

Fig. 3.20 Topology of an KNX system with 11,535 devices

(20 VDC), even when they are located some distance from the power supply (taking into account the voltage drop along the cable). And, secondly, to ensure correct bus arbitration (see Sect. 3.10.4), especially in view of detecting collisions that may occur during the growing signal propagation delay caused by the increasing cable length (see Fig. 3.21).

Fig. 3.21 Installation guidelines for KNX.TP

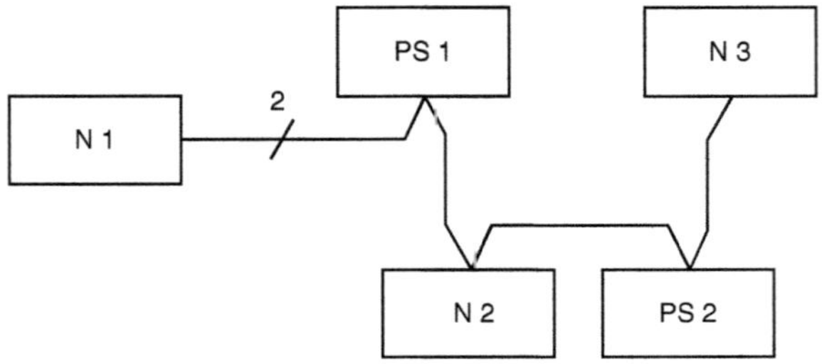

The following rules apply to a line or a line segment:

- A line must be no more than 1000 m long
- The length of cable between the two bus devices that are the furthest apart (e.g., N1 and N3 in Fig. 3.21) must be no more than 700 m
- The length of cable between a power supply unit and a device (e.g., between PS1 and N1 or between PS2 and N3 in Fig. 3.21) must be no more than 350 m
- Any two power supplies on one segment must be at least 200 m apart. More than two power supplies can not be used.

3.6.6 Block Diagrams and Standardized Device Symbols

Block diagrams with standardized symbols representing the individual devices are used to graphically display a KNX system's topology. For example, Fig. 3.22 shows the block diagram of a small KNX system. Note that you only need the

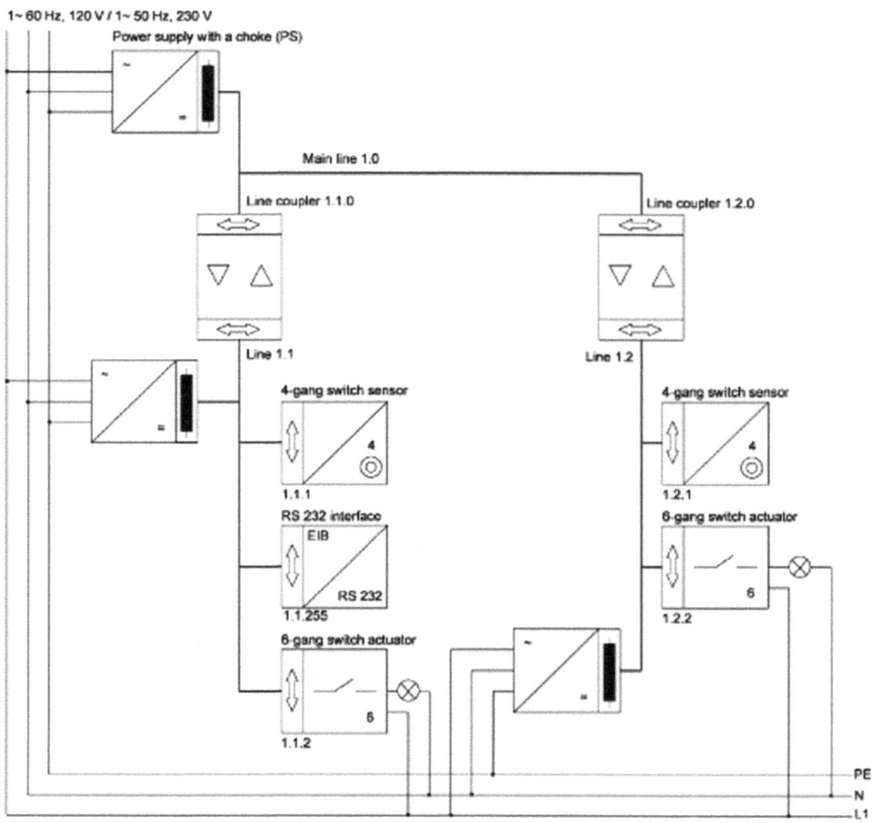

Fig. 3.22 Block diagram of a small KNX system with one area and two lines

RS-232 (serial) interface when programming (physical addresses, application programs) the bus devices using ETS 5. You can therefore remove it afterwards, because the system does not need it to operate.

3.7 Addressing Nodes (Devices)

All KNX devices that communicate (sensors, actuators, couplers, etc.) must have a unique physical address. Nodes also belong to a group so that they can exchange data with each other. The nodes in such a group have a group address in addition to their physical address, which can be used to communicate with them.

3.7.1 Physical Addresses

Each bus device (except the power supply) is assigned a unique physical address. This address is then loaded into the device (node) and stored permanently in its EEPROM during commissioning using ETS 5.

Addresses should be assigned at random, but should correspond to the layout of the building's installation. In other words, nodes (devices) that are located next to each other should be assigned consecutive physical addresses, for example, 1.1.1, 1.1.2, 1.1.3, and so on. The physical addresses must be clearly and permanently labeled on the devices.

A device's physical address clearly identifies it and, at the same time, provides the following information on the device's topographical position within the whole system:

Area.Line.Node (A.L.N.)

The area, line and node parts of the physical address are separated by periods. This way you can distinguish between a physical address and a group address, which uses a slash (/). For a physical address, 2 bytes are available in a data frame as seen in Table 3.4.

Table 3.4 Source address in a data frame

High byte								Low byte							
D7	D6	D5	D4	D3	D2	D1	D0	D7	D6	D5	D4	D3	D2	D1	D0
A3	A2	A1	A0	L3	L2	L1	L0	N7	N6	N5	N4	N3	N2	N1	N0
Area				Line				Node (per line)							

There can be

- $2^4 = 16$ areas
- $2^4 = 16$ lines
- $2^8 = 256$ nodes per line

giving a total of $2^{16} = 65{,}536$ nodes. However, this maximum capacity is rarely used. In particular, apart from couplers no other devices are connected to the main lines and the backbone line. This minimizes the delay to data transmitted across lines and areas caused by the internal data traffic from the nodes on a main line or the backbone.

The physical addresses are usually limited to:

- 15 areas (numbers 1–15)
- 12 lines per area (numbers 1–12)
- 64 nodes per line (63 sensors/actuators with numbers 1–63 and one line coupler with the node number 0)

This means that a maximum of $63 \times 12 \times 15 = 11{,}340$ sensors/actuators, plus $12 \times 15 = 180$ LCs and 15 BCs, giving a total of 11,535 nodes (Fig. 3.20).

This is more than enough for most projects in residential and commercial buildings. The majority of KNX projects usually only have a few dozen to a few hundred nodes.

3.7.1.1 Physical Addresses for Couplers and Line Repeaters

The following physical addresses are reserved for line and backbone couplers:

- A.L.0 for line couplers (e.g. 1.1.0, 1.2.0, …, 1.12.0, 2.1.0, 2.2.0, …, 15.12.0)
- A.0.0 for backbone couplers (e.g. 1.0.0, 2.0.0, …, 15.0.0)

Line repeaters must be assigned a node number that is greater than zero, for example, 1.1.64.

3.7.1.2 Physical Addresses for Main Lines and Backbone Lines

Physical Adresses of Devices Connected to a Main Line

Sensors and actuators are not normally connected to the main line. This is because the resultant data traffic would interfere with the data traffic across the lines or areas.

As a result, a main line's power supply unit usually only needs to supply one backbone coupler with power. For this reason, a maximum of 63 sensors/actuators can be connected to the main line. They are assigned the physical addresses A.0.x, where x can vary from 1 to 255. The address A.0.0 is reserved for the backbone coupler.

Examples of addresses:

- 1.0.1–1.0.63
- 2.0.100–2.0.162.

If the KNX system comprises only one area and, therefore, does not need a backbone coupler, then a maximum of 64 sensors/actuators can be connected to the main line.

Physical Adresses of Devices connected to a Backbone Line

As with the main lines, sensors and actuators are not normally connected to the backbone line. This is because the resultant data traffic would interfere with the cross-area data traffic.

As a result, a backbone line's power supply unit does not normally have to supply a bus device with power. For this reason, a maximum of 64 sensors/actuators can be connected to the backbone line. They are assigned the physical addresses 0.0.x, where x can vary from 1 to 255.

Examples of addresses:

- 0.0.1–0.0.64
- 0.0.100–0.0.163.

3.7.1.3 Examples of Physical Addresses

- The address 1.2.2 refers to the second node on the second line in the first area.
- 1.12.0 is a line coupler that connects the 12th line in the 1st area to the 1st area's main line. The main line is referred to as the superordinate line and the line is referred to as the subordinate line.
- 2.0.0 represents a backbone coupler that connects the second area's main line to the backbone line. The backbone line is the superordinate line and the main line is the subordinate line.

3.7.2 Group Addresses (Logical Addresses)

When commissioning a KNX system using ETS 5, specific devices must be addressed or programmed. For this reason, data frames are sent using a destination device's unique physical address as the destination address. During the normal operating cycle of a KNX system data frames are sent using group addresses.

KNX distinguishes between two kinds of group addresses:

- Group addresses with a main group and a subgroup (two-level addressing)
- Group addresses with a main group, middle group, and subgroup (three-level addressing)

You can use ETS 5 to select the desired addressing type. A 16-bit field in the data frame is reserved for the group address, although only 15 bits are used.

3.7.2.1 Two-Level Addressing

For two-level addressing, the available 15 bits are used as follows (Table 3.5).

This means that

- $2^4 = 16$ main groups (numbers 0–15)
- $2^{11} = 2048$ subgroups (numbers 0–2047)

are possible. The group address for two-level addressing is written as

Main group/subgroup

In order to be able to identify a certain group, these main groups and subgroups are assigned names.

Examples for group addresses:

- 0/1 lightning central on/off
- 1/1 lightning living room on/off
- 1/2 lightning office on/off
- 2/1 blinds down/up.

3.7.2.2 Three-Level Addressing

For three-level addressing, the available 15 bits are used as follows (Table 3.6).

This means that

- $2^4 = 16$ main groups (numbers 0–15)
- $2^3 = 8$ middle groups (numbers 0–7)
- $2^8 = 256$ subgroups (numbers 0–255)

are possible. The group address for three-level addressing is written as

Table 3.5 Two-level addressing

High byte								Low byte							
D7	D6	D5	D4	D3	D2	D1	D0	D7	D6	D5	D4	D3	D2	D1	D0
0	M3	M2	M1	M0	S10	S9	S8	S7	S6	S5	S4	S3	S2	S1	S0
Main group					Subgroup										

Table 3.6 Three-level addressing

High byte								Low byte							
D7	D6	D5	D4	D3	D2	D1	D0	D7	D6	D5	D4	D3	D2	D1	D0
0	M3	M2	M1	M0	G2	G1	G0	S7	S6	S5	S4	S3	S2	S1	S0
Main group					Middle group			Subgroup							

Main group/middle group/subgroup

Examples for group addresses:

- 1/1/1 lightning living room ceiling on/off
- 1/1/2 lightning living room standing lamp on/off
- 1/2/1 lightning office ceiling on/off
- 1/2/2 lightning office desk on/off.

3.7.2.3 Main Groups 14 and 15

With both two and three-level addressing, there is not enough room in the couplers' filter tables for the main groups 14 and 15 due to the limited amount of memory in the EEPROM. As a result, they are not usually included in the project design. If they are included, then the couplers must be configured accordingly.

The main group 0 is usually reserved for alarm functions, whereas the main groups 1–13 are reserved for various systems such as lighting and shutters/blinds.

3.8 Communication Objects

3.8.1 Definition

A communication object is a memory area that is used for data exchange with other applications or devices by the application software of a KNX device in combination with the communication software (see Fig. 3.23).

3.8.2 Characteristics of Communication Objects

3.8.2.1 Structure

A communication object (C.Obj.) has a certain structure. For example it can be:

- a bit field (length 1 bit, 4 bit, 8 bit etc.)
- a variable (integer, float)

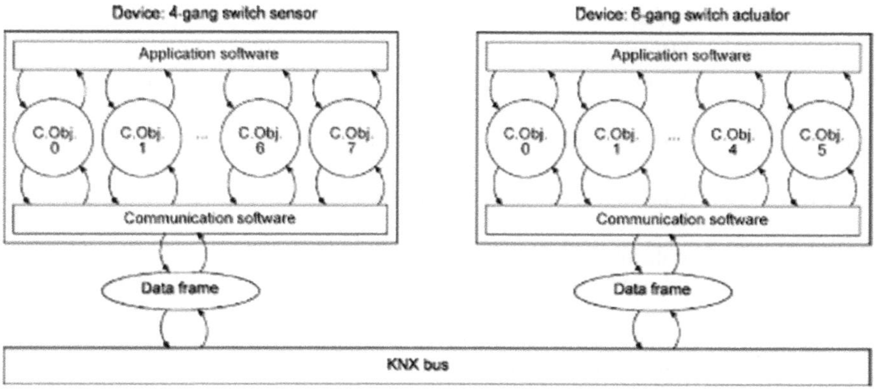

Fig. 3.23 Communication Object of a four-gang Switch Sensor and a six-gang Switch Actuator

- a time or date
- a text (e.g. 14 ASCII letters).

3.8.2.2 Attributes

For each application program there is a certain number of C.Obj. that can be found in the device's documentation. A C.Obj. has certain attributes. For example:

- number
- name
- function
- group address
- length (1 bit, 4 bit, 1 B etc.)
- flags: communication (C), read (R), write (W), transmit (T), update (U), read on initialization (I)

The flags of a C.Obj. are usually predefined (default settings) and should only be changed in special cases. For sensors and actuators, every flag besides the L-flag is usually already set. Only some devices have an I-flag.

Generally, the flags influence the access to the C.Obj. The meaning of the flags can be found in Table 3.7.

3.8.2.3 Access Methods

The C.Obj. is accessed by the ETS5, the system software and the application with certain services for reading and writing. A commonly used service in the application layer is e.g. A_GroupValue_Write, which is used to write the object value.

Table 3.7 Flags

Flag	Flag set	Flag not set
C (Communication)	C.Obj. is connected to the bus	Acknowledgement of frames but C.Obj. is not changed
R (Read)	Value of C.Obj. can be read by bus	Value of C.Obj. can not be read by bus
W (Write)	Value of C.Obj. can be changed by bus	Value of C.Obj. can not be changed by bus
T (Transmit)	If the objects value changes (for a sensor), the new value is transmitted by bus	Object value is transmitted by bus only during read commands
U (Update)	Objects value is updated by a value answer frame on the bus (service A_Value_Response). Value answer frame is a reaction to a value reading frame (service A_Value_Read) sent by e.g. a visualization	Objects value is not updated
I (Read on initialization)	C.Obj. reads the object's value from the bus during initialization (only some devices)	C.Obj. does not read the objects value during initialization

3.8.3 Communication Objects in Sensors

Sensor applications normally only use communication objects for sending data frames. This is shown by the words *Telegr.* in the function assigned to the communication object, for example, *Telegr.switch* or *Telegr. relative dimming*. As an example, Table 3.8 lists the communication objects for the *Switch Dim LED* application of a four-gang switch sensor in the *switch/dimming sensor* setting [9].

In the *switch/dimming sensor* default setting, the *Switch Dim LED* application has eight communication objects for four push buttons that are used to carry out the following functions [9]:

Table 3.8 The communication objects for the *Switch Dim LED* function in a four-gang switch sensor

No.	Type (bit)	Object name	Function
0	1	Left push button—short	Telegr.switch
1	1	Mid left push button—short	Telegr.switch
2	1	Mid right push button—short	Telegr.switch
3	1	Right push button—short	Telegr.switch
4	4	Left push button—long	Telegr. relative dimming
5	4	Mid left push button—long	Telegr. relative dimming
6	4	Mid right push button—long	Telegr. relative dimming
7	4	Right push button—long	Telegr. relative dimming

- Pressing and releasing the push buttons upper contact (the corresponding C-Obj. then contains a "1") sends a switch ON data frame. Pressing and releasing the push button's lower contact (the corresponding C.Obj. is assigned a "0") sends a switch OFF data frame.
- Pressing and holding down a push button's upper contact sends a brighter-dimming data frame. Pressing and holding the lower contact sends a darker-dimming data frame. As soon as the push button is released, a *stop dimming* data frame is sent.

3.8.4 Communication Objects in Actuators

Actuator applications normally only use communication objects to receive data frames and to execute functions. As an example, Table 3.9 shows the communication objects for the *Switch Default Staircase function* application of a six-gang switch actuator in *normal operation* mode [9].

They are used to process the following functions [9]:

- If the *Switch function* parameter has been set to *normally opened contact*, then the actuator switches the relay on when it receives a data frame with the value "1" (the corresponding C.Obj. then contains a "1") and switches it off when it receives a data frame with the value "0" (the corresponding C.Obj. then contains a "0."
- If the *Switch function* parameter has been set to *normally closed contact*, then the actuator switches the relay on when it receives a data frame with the value "0" and off when it receives a one with the value "1."

3.8.5 Assigning Communication Objects to Group Adresses

3.8.5.1 General Rules

In order to realize a function like switching or dimming, the application of at least one sensor and the application of at least one actuator have to exchange data.

Table 3.9 The communication objects for the *Switch Default Staircase function/3* in a six-gang switch actuator

No.	Type (bit)	Object name	Function
0	1	Output A	Switch
1	1	Output B	Switch
2	1	Output C	Switch
3	1	Output D	Switch
4	1	Output E	Switch
5	1	Output F	Switch

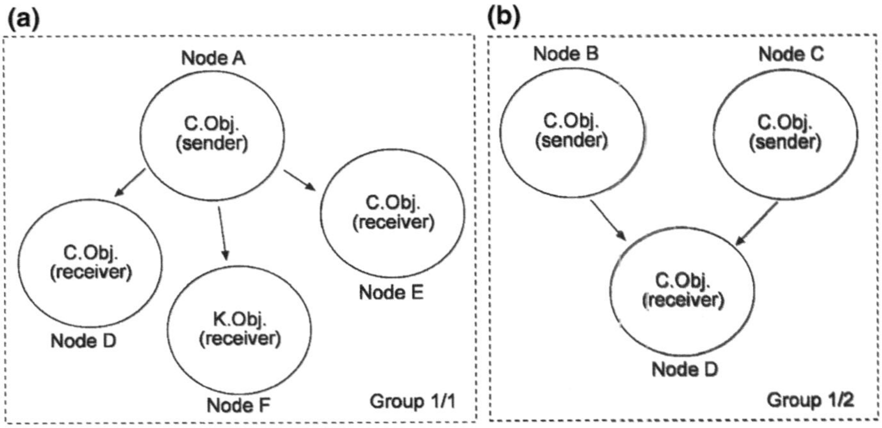

Fig. 3.24 Groups with sender and receiver communication objects

Therefore both the transmitting and the receiving applications use a number of C.Obj. that have to share one group address. This is equivalent to a connection of the devices.

Assuming that nodes A, B and C are sensors and that nodes D, E and F are actuators with the respective C.Obj. as seen in Fig. 3.24.

In order to realize two different functions with ETS5, you need to create group addresses (1/1 and 1/2) that will be used to carry out the desired functions. The communication objects for all the bus devices required for the function must then be assigned to these group addresses. This can be done by dragging and dropping. Also note that the communication objects that are to be assigned together must all be the same type (length), for example, all 1 bit, 4 bit or 1 B. ETS 5 will not allow you to assign them together if they are not all of the same type.

In general, the following rules apply:

- A group address needs to include at least one transmitting and at least one receiving C.Obj.
- A transmitting C.Obj. can only be assigned to one group address.
- A receiving C.Obj. can be assigned to multiple group addresses.

Like the number of communication objects that can be assigned to a group address, there are only a limited number of communication objects and group addresses that can be used by an application. This information can be found in the technical data documentation that comes with each KNX device. An example is shown in Fig. 3.25. The number of communication objects varies depending on the application and the parameters that have been set. The number of functions assigned to the communication objects also varies from application to application.

ABB i-bus® EIB — Switch Actuator, 6-fold, 10 A, MDRC
AT/S 6.6.1, GH Q631 0023 R0111

Application programs	Number of communication objects	Max. number of group addresses	Max. number of associations
Switch Logic Default /3	12	16	17
Switch Status Default /1	12	18	18
Switch Default Time /3	6	18	18
Switch Default Stairc.fct /3	6	21	22
Switch Priority Status Default /6	12	14	14
Fan coil 4-pipe Heat and Cool /1	5	5	5

Fig. 3.25 Application programs with the number of communication objects, group addresses and associations [9]

3.8.5.2 Example

The following function is to be executed using KNX: The left push button of a four-gang switch sensor is to turn two lamp circuits on and off. Pressing the push button's upper contact should switch all the lamps on and pressing the lower contact should turn them all off. A six-gang switch actuator's outputs A and B will switch the lamp circuits.

In order to know, which C.Obj. are available, the correct applications have to be selected for 4-gang switch sensor and 6-gang switch actuator. It is also possible that setting the application parameters results in different C.Obj.

Therefore, the following steps are to be taken:

1. Select application
2. Set parameters of the application
3. Create group addresses
4. Assign communication objects to group addresses

Figure 3.26 shows how the two devices are connected. C.Obj. 0 (object name: *left push button—short*) is assigned to the *left push button*, C.Obj. 0 (object name: *Output A*) of the six-gang switch actuator is assigned to output (relay) A, and C.Obj. 1 (object name: *Output B*) of the six-gang switch actuator is assigned to output (relay) B. All C.Obj. are assigned to the same group address 1/1. After programming both devices, the system is ready for use.

Pressing the upper contact of the left push button sends a Switch ON data frame (with group address 1/1, the command *write* value, and the object value "1"), which closes the relay contacts for outputs A and B and turns on the lamps. And in same way, pressing the lower contact of the left push button, sends a Switch OFF data frame (with the group address 1/1, the command *write* value, and the object value "0"), which then opens the relay contact for outputs A and B and turns off the lamps.

This is assuming that the *Switch function* parameters for outputs A and B have both been set to *normally opened contact*, and also that the communication flag and the transmission has been set for C.Obj. 0 and 1 in the switch actuator. The C.Obj. then has a standard connection to the bus, which if the object value changes, a data

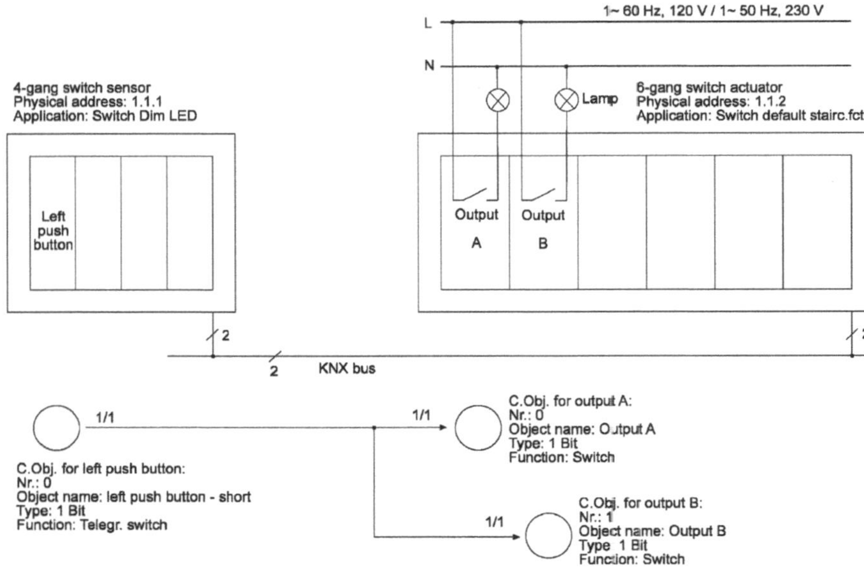

Fig. 3.26 Assigning communication objects to a group address

frame is sent. The communication flag and the write flag must be set for C.Obj. 0 and 1 of the switch actuator. This way the C.Obj. also has a normal connection to the bus, which means that the object value can be changed via the bus.

3.9 User Data

A KNX data frame can contain between two and 16 bytes of user data (there are also expanded data frames with up to 255 bytes of user data). During normal operation, user data is mainly two bytes long (usually switching or dimming commands).

3.9.1 Accessing Services of the Application Layer

Ten bits in the first two bytes of the user data of a data frame are reserved for services that the application layer of the application provides. However, commonly used services like e.g.

- Reading a C.Obj.: A_GroupValue_Read (A_ stands for application layer) and
- Writing a C.Obj.: A_GroupValue_Write

are coded with only 4 bits, so that the remaining 6 bits can either be filled with zeros or with "actual" data as seen for switching and dimming commands.

3.9.2 *EIB Interworking Standard (EIS)*

The number of data bytes depends on the predefined EIB Interworking Standards (EIS) types, which enable specific functions to be carried out irrespective of the manufacturer [8]. Table 3.10 shows several examples of EIS types.

As an example for the use of EIS. the usage of the EIS types 1 (switching) and 2 (dimming) is explained further below.

3.9.2.1 EIS-Type 1: Switching

EIS-Type 1 is used for the function switching. The code of a switching command includes 5 bits and is stored in two user data bytes (byte 6 and byte 7 of a standard data frame) as seen in Table 3.11.

The meaning of the particular bits is as follows:

- Bits B9, B8, B7 and B6 contain 0010 corresponding to the service A_GroupValue_Write—writing in a C.Obj.
- Bit D0 = B0 in Byte 7 is used for the actual switching information. Value 1 means switching on and value 0 means switching off
- Every other bit is transmitted as a zero bit and not further processed.

Table 3.10 EIS types (examples)

EIS type	Function	Length of C.Obj.	Length of user data
1	Switching	1 bit	2 byte
2	Dimming	4 bits	2 bytes
3	Time (d, h, min, s)	3 bytes	5 bytes
4	Date (day, month, year)	3 bytes	5 bytes
5	Float	2 bytes	4 bytes
15	String	14 bytes	16 bytes

Table 3.11 The first two user data bytes for a switching command

Byte 6								Byte 7								Command
MSB							LSB	MSB							LSB	
D7	D6	D5	D4	D3	D2	D1 B9	D0 B8	D7 B7	D6 B6	D5 B5	D4 B4	D3 B3	D2 B2	D1 B1	D0 B0	
0	0	0	0	0	0	0	0	1	0	0	0	0	0	0	1	Switch on
0	0	0	0	0	0	0	0	1	0	0	0	0	0	0	0	Switch off

3.9.2.2 EIS-Type 2: Dimming

EIS-Type 2 is used for the function dimming. The code of a switching command includes 8 bits and is stored in two user data bytes (byte 6 and byte 7 of a standard data frame) as seen in Table 3.12.

The meaning of the particular bits is as follows:

- Bits B9, B8, B7 and B6 contain 0010 corresponding to the service A_GroupValue_Write—writing in a C.Obj.
- The actual dimming function requires 4 bits in byte 7 (D3 = B3, D2 = B2, D1 = B1 and D0 = B0)
  - Bit D3 contains the dimming information. A one bit means *increase brightness* and a zero bit means *decrease brightness*
  - Bits D2, D1 and D0 contain 011 corresponding to "dimming level 4", this means "to the next dimming level (0, 25, 50, 75, 100%) depending on the current level
- Every other bit is transmitted as a zero bit and not further processed.

3.9.3 Length of User Data

The number of user data bytes is explicitly stated in byte 5 of the standard data frame, so that the receiver can act accordingly (Table 3.13).

That means that the decimal value of the length value increased by 1 is the actual number of user data bytes in a standard data frame.

3.10 The Communication Process

A data frame is sent in response to an individual action such as pressing the upper-left rocker switch on a four-gang switch sensor (there are also KNX devices that send data frames periodically). The bus device then sends a data frame that has a specific group address.

All devices (receivers) that belong to this group simultaneously confirm that they have received the data frame by returning an acknowledgment frame. This acknowledgment frame is also called a summation frame, because it comprises the confirmation frames from all the receivers. If the sender transmits a frame to a device located on another line, the coupler confirms receipt of the frame.

This section will explain:

- The types of data frames used in KNX
- How a data frame arrives on the bus
- How a data frame is sent (as a sequence of UART characters)

Table 3.12 The first two user data bytes for a switching command

Byte 6								Byte 7								Command
MSB							LSB	MSB							LSB	
D7	D6	D5	D4	D3	D2	D1 B9	D0 B8	D7 B7	D6 B6	D5 B5	D4 B4	D3 B3	D2 B2	D1 B1	D0 B0	
0	0	0	0	0	0	0	0	1	0	0	0	1	0	1	1	Dimming

Table 3.13 Length of L3, L2, L1, L0

D7	D6	D5	D4	D3	D2	D1	D0	
DAF	R2	R1	R0	L3	L2	L1	L0	
				0	0	0	1	2 user data bytes
				⋮	⋮	⋮	⋮	⋮
				1	1	1	1	16 user data bytes

- How a receiver is addressed
- How long it takes for a sender and a group of receiver(s) to communicate with each other.

3.10.1 Frame Types

Information is exchanged between KNX devices by using frames. There are three types of frames: data frames, poll data frame and acknowledgment frames.

3.10.1.1 Data Frames

There are two types of data frames:

- Standard data frame (short data frame, up to 16 user data bytes)
- Expanded data frame (long data frame, up to 255 user data bytes).

3.10.1.2 Poll Data Frame

With the poll data frame one node (master) can request a 1-byte-information from up to 15 bus devices (slaves). The bus devices are addressed using a group address. The frame consists of the poll data request by the master and the poll data response by the slaves. Every slave stores its information in a certain time slot. Group address and number of this slot can be set using ETS 5 as well as the one byte information to be send by every slave. All values from 0x00 up to 0xFF are valid apart from 0xFE. A poll data frame can for example be used to check if certain bus devices are working.

3.10.1.3 Acknowledgment Frame

An acknowledgment frame consists of only one byte—the control field. The bits stored there are explained in Sect. 3.10.7.

3.10.1.4 Type of Frame in the Control Field

The first byte of every data or poll data frame (byte 0) is a control field. Its bits D7 and D6 define the type of the frame as seen in Table 3.14.

The R-bit (repeat flag) and the two priority bits (P1 and P0) will be explained later on. To understand the fundamental processes of the communication in a KNX system, working with the standard data frame is sufficient.

3.10.2 Structure of a Standard Data Frame

The information in a standard data frame is structured in eight bit fields, see Table 3.15.

The shortest possible standard data frame has a length of 9 bytes (byte 0 to byte 8 with 2 user data bytes), the longest has a length of 23 bytes (byte 0 to byte 22 with 16 user data bytes).

3.10.3 Universal Asynchronous Receive Transmit (UART)

A KNX data frame is a series of UART characters.

3.10.3.1 UART Characters

A sequence of UART characters contains the following bits:

- 1 start bit (SB, with 0 value)
- 8 data bits (D7, D6, …, D0)

Table 3.14 Control field (Byte 0 of a frame)

D7	D6	D5	D4	D3	D2	D1	D0	
1	0	R	1	P1	P0	0	0	Standard data frame
0	0	R	1	P1	P0	0	0	Expanded data frame
1	1	1	1	0	0	0	0	Poll data frame

Table 3.15 Bit fields in a standard data frame

1 Byte	2 Bytes	2 Bytes	1 Byte			2 bis 16 Bytes	1 Byte
			1 Bit	3 Bits	4 Bits		
Control field	Source address	Destination address	Address type	Routing counter	Length of user data	User data	Test field

- 1 parity bit (PB, even parity)
- 1 stop bit (or end bit) (EB, with 1 value)

After a UART character has been sent, the sender waits for 2 bit times to elapse before it sends the next character. As a result, it takes a total of 13 bit times to send a KNX character sequence. There is an interval of one character between the last UART character of a data frame and the acknowledgment (summation) frame sent by the receivers. This means that an interval of 15 bit times elapses between the stop bit of the last sequence of a data frame and the start bit of the acknowledgment frame.

3.10.3.2 Sending UART Characters

Assuming that a data frame's control field (control byte) contains the following bits (Table 3.16).

The bit on the far left is the most significant bit (MSB) and the one on the far right is the least significant bit (LSB). To send the control field as a sequence of UART characters, the UART module automatically adds a start bit (SB), a parity bit (PB) with even parity, and an end bit (EB) to the data bits, and reverses the order of the data bits D7–D0 (see Table 3.17).

After a UART character has been sent, the sender waits for 2 bit times.

3.10.4 Bus Arbitration

When a KNX devices sends a frame, it first checks if the bus is idle and starts transmitting if this is the case. However, if two or more nodes try to send at the same time the result is a bus access conflict. In order to avoid these conflicts and to define which node is allowed to send first (Bus Arbitration), KNX uses two different concepts:

- Access Classes
- CSMA/CA.

Table 3.16 Control field bits

MSB							LSB
D7	D6	D5	D4	D3	D2	D1	D0
1	0	1	1	1	1	0	0

Table 3.17 Actual sequence of bits when sent as a sequence of UART characters

SB	D0	D1	D2	D3	D4	D5	D6	D7	PB	EB
0	0	0	1	1	1	1	0	1	1	1

3.10.4.1 Idle Bus

If a device wants to send a data frame over the bus, it must first make sure that no data is being transmitted, in other words, that the bus is free. If a device needs to send a frame, it listens to see if there is any traffic on the bus. If no data is transmitted during the time it takes to send 50 bits (50 bit times), then the bus is free (bus idle). The signal is the same as the bus signal for a sequence of 50 “one” bits, which means the difference in voltage between the red and black wires is always (nominally) 24 V DC.

3.10.4.2 Access Classes

In order to reduce the number of potential bus access conflicts, special access classes have been defined. Access class 1 data frames can be sent if there has been no activity on the bus for a period of 50 bit times. Access class 2 data frames, on the other hand, are only allowed to be sent 3 bit times later—or 53 bit times without any bus activity. If an access class 1 data frame is sent, the bus is no longer free for access class 2 data frames.

Frames belonging to access class 1:

- System frames (system priority)
- Alarm frames (alarm priority)
- Repeated data frames

Frames belonging to access class 2:

- Priority frames (high priority)
- Normal frames (low priority)

The priority of a data frame is defined in the control field, see Sect. 3.10.4.4.

If an access class 1 data frame is sent, the bus is no longer free for access class 2 data frames. They can only be transmitted when all access class 1 data frames have been sent. The CSMA/CA algorithm is used when bus access conflicts occur between data frames in the same access class (access class 1 or 2).

3.10.4.3 Carrier Sense Multiple Access/Collision Avoidance

CSMA/CA is an algorithm that can resolve bus access conflicts, enabling one sender to send its data frame without a delay (Fig. 3.28).

Dominant and Recessive Bits

The CSMA/CA is a stochastic media access control protocol and requires both dominant and recessive signal elements to be present on the bus. This means that the dominant signal element has priority over the recessive signal element. In KNX the signal element assigned to the "zero" bit is dominant, and the signal element assigned to the "one" bit is recessive.

As this book is only meant to be an introduction, we will not go into further detail into how CSMA/CA is implemented in KNX. For further information, see [8].

We will, however, still explain one method that uses dominant and recessive bits —Wired-AND Switching (see Fig. 3.27).

Three transistors (with basis B, collector C, and emitter E) are used as electrical switches in a wired-AND switching circuit. If the basis-emitter voltage is $U_{BE} \leq 0$, the current cannot flow from collector C to emitter E (open contact). If $U_{BE} > 0$, then the current flows—C and E have practically the same potential: contact is open. If $U_{BE} \leq 0$ for all (wired-AND) transistors, a current cannot flow through the pull-up resistor, and then: $U_{Bus} = 0$. As soon as only one transistor's basis-emitter voltage is higher than zero (contact is closed), the potential on the collector is drawn to the emitter's potential and U_{Bus} becomes zero. Therefore $U_{BE} > 0$ represents the dominant bit.

Bit-by-bit bus arbitration can be carried out using CSMA/CA. One device always has priority over the other devices, and it is allowed to send all its data frames without any time delay.

Process of CSMA/CA

The following example is an analogy of how the CSMA/CA works (Fig. 3.28).

Two people are sitting in front of a window in a dark room at night. They both want to send a message ("frame") to someone outside using their flashlights. They agree the following signals ("signal elements"):

- "Zero" bit: flashlight is turned on for one second (room is lit)
- "One" bit: flashlight is turned off for one second (room is dark).

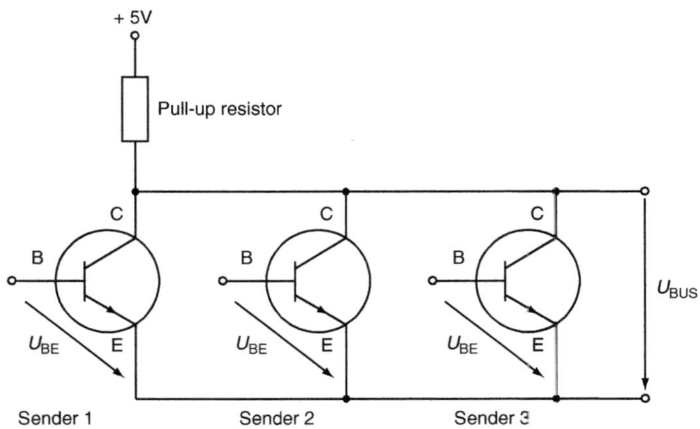

Fig. 3.27 Wired-AND switching: dominant and recessive bits

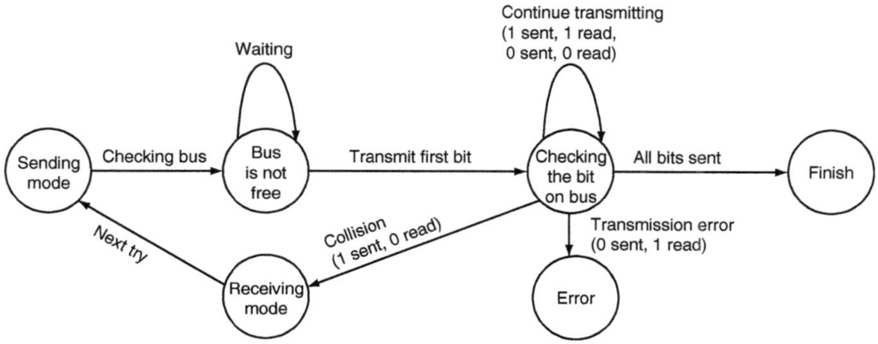

Fig. 3.28 Status diagram for the CSMA/CA protocol

The bit time is defined as 1 s, which means the bit rate is 60 bit/s. The following bit sequences are to be sent:

- Person A: 0 1 1 0 (light–dark–dark–light),
- Person B: 0 1 0 1 (light–dark–light–dark).

If the room is dark (no bus activity) for a period of 50 bit times (here: 50 s), then the two people start to send their messages by turning on their flashlights for one second ("zero" bit, light). They both see the "light" signal element, which corresponds to the signals they are sending, and they therefore both continue sending.

The second bit ("one" bit, dark) also corresponds to the "dark" signal element they are both sending. Both continue sending.

A collision occurs, however, when the third bit is sent. The dominant "zero" bit (light) sent by person B prevails—the room is lit even though person A wants it to be dark. So person A loses the bus arbitration, stops sending and changes to receipt mode. Person B wins the bus arbitration and can therefore continue sending the remaining bits of his or her message. Person A tries again after person B has finished transmitting.

3.10.4.4 Criteria for Bus Access

Bus arbitration in KNX is resolved using:

- Two parity bits D3 (P1) and D2 (P0) in the control field
- The repeat flag bit D5 (R) in the control field
- The source address bits

Access classes also influence the order in which the frames are sent (even before the CSMA/CA method begins).

Priority Types

Table 3.18 shows the priority types used in KNX. A system data frame has the highest priority and a normal data frame has the lowest priority.

The P0 priority bit is the most important bit in bus arbitration, because it is sent before the P1 priority and the repeat flag R bits. This is because the frame that wins access to the bus is the one that is the first to have a dominant bit when the other has a recessive bit. Most data frames are normal data frames with low priority. If data frames have the same priority, the repeat flag bit can be used for arbitration.

Repeat Flag Bits

When a data frame is sent for the first time, the sender may receive a negative acknowledgment (NACK) from one or more of the receivers (see Sect. 3.10.7.1). The sender must then send the data frame again. The repeat flag bit in the repeated data frame is set to zero so that the receivers that have already successfully received and processed this data frame do not process it again.

Repeated data frames belong to access class 1 and can be send 3 bit times earlier than data frames that are sent for the first time. Repeated data frames can cause conflict with system frames and alarm frames and this is where the repeat flag bit becomes important for the decision. There are the following possibilities (Table 3.19).

System frames and alarm frames always have a higher priority than repeated standard data frames. Repeated system frames always have a higher priority than repeated alarm frames.

Source Addresses

If data frames have the same priority bit and repeat flag bit, bus arbitration is carried out using the (unique) source address. Since bus arbitration only takes place within one line or one line segment (the couplers and line repeaters represent the boundary

Table 3.18 Priority types

D3 (P1)	D2 (P0)	Priority (frame type)
0	0	System priority (system frame)
1	0	Alarm priority (alarm frame)
0	1	High priority (priority frame)
1	1	Low priority (normal frame)

Table 3.19 Repeated flag bit for bus arbitration

D5 R	D3 P1	D2 P0	Frames of access class 1
1	0	0	Frame with system priority
1	1	0	Frame with alarm priority
0	0	0	Repeated data frame with system priority
0	1	0	Repeated data frame with alarm priority
0	0	1	Repeated data frame with high priority
0	1	1	Repeated data frame with low priority

of bus arbitration), the bus access conflict is resolved using the device number. The sender that wins is not always the one with the smallest device number, but rather the one that—when the sequence of UART characters is compared bit by bit—has a "zero" bit when the others have a "one" bit.

3.10.4.5 Example of Bus Arbitration

Four nodes want to send data frames. Until these data frames have been processed, no other nodes may transmit data. Table 3.20 shows the fields of the frame used in bus arbitration.

Access Class

The access class is the first criterion to be considered in bus arbitration. Node 1.1.1 wants to send a low-priority data frame for the first time and node 1.1.2 wants to send a high-priority data frame for the first time. These data frames both belong to access class 2 and can therefore only be transmitted after 53 bit times. Nodes 1.1.11 and 1.1.42, on the other hand, want to re-send their data frames—which means they belong to access class 1 and can therefore be transmitted after 50 bit times. This leads, however, to a bus access conflict between nodes 1.1.11 and 1.1.42, which is resolved with CSMA/CA.

CSMA/CA

Bus arbitration only takes into account the data bits of the UART characters. Table 3.21 shows all bits as they will occur on the bus by time.

Comparing the first sequence of UART characters (control field) bit by bit from left to right, it can be noticed that they are identical: the priority bits and repeat flag bits are the same. All the bits in the second sequence of UART characters (the more significant byte of the source address with the area and line number) are also the

Table 3.20 Fields used in bus arbitration: nodes 1.1.1, 1.1.2, 1.1.11, and 1.1.42

Node	Control field								Source address		
	D7	D6	R	D4	P1	P0	D1	D0	Area	Line	Node
1.1.1	1	0	1	1	1	1	0	0	0001	0001	00000001
1.1.2	1	0	1	1	0	1	0	0	0001	0001	00000010
1.1.11	1	0	0	1	0	1	0	0	0001	0001	00001011
1.1.42	1	0	0	1	0	1	0	0	0001	0001	00101010

Table 3.21 Fields used in bus arbitration: nodes 1.1.11 and 1.1.42

Node	UART characters 1 +2 pause bits	UART characters 2 +2 pause bits	UART characters 3 +2 pause bits
1.1.11	0 00101001 11 11	0 10001000 01 11	0 11010000 11 11
1.1.42	0 00101001 11 11	0 10001000 01 11	0 01010100 11 11
Time	→		

Table 3.22 Fields used in bus arbitration: nodes 1.1.1 and 1.1.2

Node	UART characters 1 +2 pause bits	UART characters 2 +2 pause bits	UART characters 3 +2 pause bits
1.1.1	0 00111101 11 11	0 10001000 01 11	0 10000000 11 11
1.1.2	0 00101101 01 11	0 10001000 01 11	0 01000000 11 11
Time	→		

same. It is only when we compare the third sequence of UART characters (the less significant byte of the source address with the node number) that we notice a difference: the first data bit sent by node 1.1.42 is a “zero” bit (dominant), whereas the first data bit sent by node 1.1.11 is a “one” bit (recessive). Node 1.1.42 therefore has priority over node 1.1.11 and the bus conflict is resolved: node 1.1.11 must stop sending its data frame. Node 1.1.42 can continue sending the remaining UART characters of its data frame.

Once node 1.1.42 has finished sending, node 1.1.11 can then send its data frame next, because it is the only remaining frame belonging to access class 1. Another bus conflict now occurs, however, between nodes 1.1.1 and 1.1.2 (see Table 3.22).

Comparing the first sequence of UART characters bit by bit from left to right, it can be noticed that node 1.1.1 wants to send a low-priority data frame (two “one” bits at position 4 and 5), whereas node 1.1.2 wants to send a high-priority data frame. Node 1.1.2 is therefore allowed to send its frame first; then node 1.1.1 is allowed to send its frame.

3.10.5 *Forwarding Data Frames*

As long as a group address contains only C.Obj. of the nodes of one line, forwarding data frames to other lines or even other areas is not necessary. However, if this is not the case, couplers are used.

3.10.5.1 Filter Function of Coupers and Line Repeaters

Line couplers and backbone couplers can operate as filters for data communication. This means, that they will only forward data frames, if the receiver of the frame is not within the same line as the transmitter. In order to be able to do so, couplers work with a filter table with group addresses. These tables are stored in the couplers permanently during initialization using ETS 5. The couplers are also equipped with a frame buffer for e.g. 50 frames each for subordinate and superior line.

Filtering data frames ensures, that they are only forwarded to where they are actually required. Data frame traffic for the entire system is reduced and the data traffic in one line is separated from that in other lines. The result is, that there can be data traffic in multiple lines at the same time. The forwarding behavior of couplers can be set using ETS 5. For example, group frames sent to the main line from one line can be

- always forwarded
- filtered or
- blocked.

Line repeaters do not work as filters. They only reconstruct the received signal and always forward it to the superior or subordinate segment. If there are errors during transmission, the forwarding of the frame can be repeated up to three times.

3.10.5.2 Routing Counter

Standard data frames can be forwarded by line repeaters (LR) and couplers (LC or BC) up to six times. The sixth byte (byte 5) of the data frame contains a 3-bit routing counter (R2, R1, R0) that controls the number of times a frame is forwarded (see Table 3.23).

The sender sets the routing counter in the normal data frame to the value six. If an LR, LC or a BC forwards the data frame, the value of the routing counter is reduced by one. Once the routing counter reaches zero the data frame is no longer forwarded.

3.10.6 Backup

Transmission errors are always a possibility during digital data transmission. Such an error is present, if a zero bit is sent but a one bit is received or vice versa.

Table 3.23 Routing counter R2, R1, R0 (in the 6th byte—byte 5—of the data frame)

D7	D6	D5	D4	D3	D2	D1	D0	
DAF	R2	R1	R0	L3	L2	L1	L0	Forwarding
	1	1	1					Unlimited
	1	1	0					Forward up to 6 times
	1	0	1					Forward up to 5 times
	1	0	0					Forward up to 4 times
	0	1	1					Forward up to 3 times
	0	1	0					Forward up to 2 times
	0	0	1					Forward up to 1 time
	0	0	0					Do not forward

3.10.6.1 Cross Parity

In KNX, block parity is used to detect transmission errors (also known as cross-parity). This method detects one, two and three-bit errors, which is sufficient for use in building control. If an error is detected, the data frame is re-sent. Error correction is not carried out.

Block parity is based on a double-parity check:

- The data byte of each sequence of UART characters is secured with even parity.
- The data frame also has one more sequence of UART characters with a parity byte. Its bits are arranged in such a way that the data and parity bits of all other data bytes are secured column by column using odd parity.

3.10.6.2 Example

The following example in Table 3.24 shows how cross parity works.

3.10.7 Acknowledgment Frames

When a data frame is received, all nodes in the addressed group wait 15 bit times and then send their acknowledgment frames at the same time. This is called a summation frame. Dominant "zero" bits overwrite the recessive "one" bits like in bus arbitration.

Table 3.24 Example of block parity in KNX

Field	UART characters			
	Start bit	Data byte (D7 … D0)	Parity bit (even)	Stop bit
Control field	0	10111100	→1	1
Source address	0	00101010	→1	1
Source address	0	00100010	→0	1
Destination address	0	01001010	→1	1
Destination address	0	00101010	→1	1
DAF/routing/length	0	11100001	→0	1
User data	0	00000000	→0	1
User data	0	10000001	→0	1
Checksum field	0	↓↓↓↓↓↓↓↓	→0	1
		01001011 (odd parity)		

3.10.7.1 The Content of an Acknowledgment Frame

An acknowledgment frame comprises only one byte and can contain the bits (cases 1–4), see Table 3.25.

Definition:

- ACK: The data frame has been received correctly (positive acknowledgment)
- BUSY: The receiver cannot process the received frame (negative acknowledgment)
- NACK: The data frame has not been received correctly (negative acknowledgment).

3.10.7.2 How a Sender Responds to an Acknowledgment Frame

If the sender detects that the acknowledgement frame is an ACK, then the sender's data has been successfully transmitted.

If the sender detects a BUSY and/or a NACK, it re-sends the data frame (as a repeated data frame) a maximum of three times.

No nodes respond if a data frame has been incorrectly addressed, for example, to a node that does not exist or if the frame has been damaged by interference voltage during transmission so that it can longer be read. The sender waits 13 bits times (bus idle) and then, as in the previous example, re-sends the data frame (as a repeated data frame) a maximum of three times.

3.10.7.3 Example 1

The three nodes that receive the data frame simultaneously send their acknowledgement frames over the bus. Table 3.26 shows the UART characters (synchronous bits) sent. The parity bit (even parity) is "zero" in all three frames.

As the receiver of the acknowledgement frame, the node that sent the data frame monitors the bus and registers the following answer (summation frame), because the "zero" bits overlay the "one" bits (Table 3.27).

Table 3.25 An acknowledgment frame (with N1, N0: NACK bits; B1, B0: BUSY bits)

	MSB						LSB	
	D7 N1	D6 N0	D5	D4	D3 B1	D2 B0	D1	D0	
Case 1	1	1	0	0	1	1	0	0	ACK
Case 2	1	1	0	0	0	0	0	0	BUSY
Case 3	0	0	0	0	1	1	0	0	NACK
Case 4	0	0	0	0	0	0	0	0	NACK + BUSY

Table 3.26 The three receiver's acknowledgement frames

Receiver	UART characters	Definition
1	0 00110011 01	ACK
2	0 00000011 01	BUSY
3	0 00110000 01	NACK

Table 3.27 The summation frame from the three receivers

UART characters (on the bus)	Definition
0 00000000 01	BUSY + NACK

For the sender (only the data bits of UART characters are interpreted), this means that at least one receiver was busy (BUSY—D3, D2 are "zero" bits) and at least one receiver received a corrupted data frame (NACK—D7, D6 are "zero" bits).

The sender cannot detect whether receiver 1 received the data frame correctly and returned an ACK, because the dominant BUSY and NACK "zero" bits overwrite any recessive "one" bits.

3.10.7.4 Example 2

Three receivers of a data frame have the physical addresses 1.1.10, 1.1.30 and 1.2.1, the transmitter is 1.1.1. C.Obj. of receiver and transmitter required for the function are assigned to the group address 1/1. Until all the receivers received the information from the transmitter, three data transmissions (data frame and acknowledgment frame) are required:

1. Transmitter 1.1.1 sends a data frame on line 1.1, receiver 1.1.10, 1.1.30 and line coupler 1.1.0 send a summation frame as acknowledgment
2. Line coupler 1.1.0 sends a data frame on line 1.0 (main line) that is acknowledged by line coupler 1.2.0
3. Line coupler 1.2.0 sends a data frame on line 1.2 that is acknowledged by receiver 1.2.1.

3.10.8 The Length of the Communication Process

Many data frames—especially those with switching commands—contain only two data bytes. In this case, it takes a total of 20.072 ms for a sender and a receiver to communicate with each other (Table 3.28).

The interval after the data frame is either 13 or 15 bit times. This depends on whether the two interval bits after the data frame's last UART character (with the checksum field) are added to the UART characters or not. Either way, there is an

Table 3.28 The time taken to transmit data and acknowledgement frames

Activity	Duration in bit times	Duration in μs
Waiting for free bus	53	5512
Sending data frame	$8 \cdot (11 + 2) + 11 = 115$	12,168
Interval	15	1352
Sending acknowledgment frame	11	1144
Sum	194	20,176

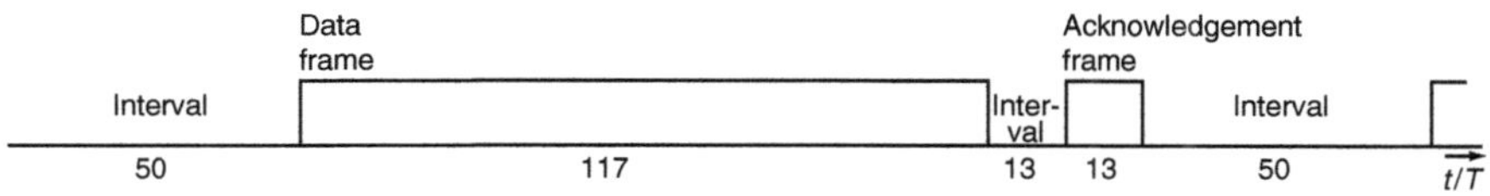

Fig. 3.29 The time it takes to exchange data (note for the lecture: this picture still has to be adapted to the contents of Table 3.28: 53–115–15–11–50)

interval of 15 bit times between the data frame's last stop bit and the start bit of the acknowledgment frame (15 "one" bits).

Figure 3.29 shows how long the communication process takes. The bit times are shown under the time axis.

If the data frame belongs to access class two, the interval (usually at least 50 bit times) after the acknowledgement frame is extended by three bit times (0.312 ms). This prevents many bus access conflicts right from the start. It takes ~20 ms for a sender and receiver to exchange a short data frame such as a switching command. This means that a sender can send ~50 switching commands per second. If the data frames contain the maximum number of bytes, then the data exchange will take up to 40 ms.

If a receiver is located on another line or even in another area, then each coupler has to forward the data frame to the superordinate or subordinate line. The total transmission time is then proportionally longer.

If the switch sensor 1.1.1 sends a data frame to the switch actuator 2.1.1, the data frame will pass through two line couplers and two backbone couplers (1.1.0, 1.0.0, 2.0.0 and 2.1.0). Each coupler receives the data frame, buffers it, checks its filter table to see whether it is allowed to forward it, and if so, forwards the frame to the next coupler or ultimately to the switch actuator 2.1.1.

1.1.1 → 1.1.0, 1.1.0 → 1.0.0, 1.0.0 → 2.0.0, 2.0.0 → 2.1.0, 2.1.0 → 2.1.1.

Five data transmissions must take place, including any re-sent data frames, for collisions or transmission errors to occur on a line. By the time the switch actuator 2.1.1 actually executes the switching command, almost 0.1 s will have elapsed (five times ~20 ms to transmit the data frame). Nevertheless, this response time is still sufficient for building control.

3.11 Summary of Data Frame Structures

The following gives a short summary of the structure of data frames and acknowledgment frames.

3.11.1 Standard Data Frame

See Table 3.29.

3.11.1.1 Control Field (Byte 0)

See Table 3.30.

3.11.1.2 Source Address (Byte 1 and Byte 2)

See Tables 3.31, 3.32 and 3.33.

Table 3.29 Bit fields for a standard data frame

1 Byte	2 Bytes	2 Bytes	1 Byte			2–16 Bytes	1 Byte
			1 Bit	3 Bits	4 Bits		
Control field	Source address	Destination address	DAF	Routing counter	User data length	User data	Checksum

Table 3.30 Control field

Byte 0								
D7	D6	D5	D4	D3	D2	D1	D0	
		R	1	P1	P0	0	0	
				0	0			System priority
				1	0			Alarm priority
				0	1			High priority
				1	1			Low priority
		0						Repetition
		1						No repetition
1	0							Standard data frame

Table 3.31 Source address

High byte (byte 1)								Low byte (byte 2)							
D7	D6	D5	D4	D3	D2	D1	D0	D7	D6	D5	D4	D3	D2	D1	D0
A3	A2	A1	A0	L3	L2	L1	L0	N7	N6	N5	N4	N3	N2	N1	N0
Area				Line				Node (per line)							

Table 3.32 Group address (two-level addressing)

High byte								Low byte							
D7	D6	D5	D4	D3	D2	D1	D0	D7	D6	D5	D4	D3	D2	D1	D0
0	M3	M2	M1	M0	S10	S9	S8	S7	S6	S5	S4	S3	S2	S1	S0
	Main group				Subgroup										

Table 3.33 Group address (three-level addressing)

High byte								Low byte							
D7	D6	D5	D4	D3	D2	D1	D0	D7	D6	D5	D4	D3	D2	D1	D0
0	M3	M2	M1	M0	G2	G1	G0	S7	S6	S5	S4	S3	S2	S1	S0
	Main group				Middle group			Subgroup							

3.11.1.3 Destination Address Field (DAF), Routing Counter, Length (Byte 5)

See Table 3.34.

3.11.1.4 User Data (Byte 6 and Byte 7)

See Tables 3.35 and 3.36.

3.11.1.5 Checksum (Byte 8)

See Table 3.37.

3.11.2 Acknowledgment Frame

See Table 3.38.

Table 3.34 DAF, Routing counter (R1, R2, R0), user data length (L3, L2, L1, L0)

Byte 5								
D7	D6	D5	D4	D3	D2	D1	D0	
DAF	R2	R1	R0	L3	L2	L1	L0	
				0	0	0	1	2 User data bytes
				0	0	1	0	3 User data bytes
				0	0	1	1	4 User data bytes
				0	1	0	0	5 User data bytes
				0	1	0	1	6 User data bytes
				0	1	1	0	7 User data bytes
				0	1	1	1	8 User data bytes
				1	0	0	0	9 User data bytes
				1	0	0	1	10 User data bytes
				1	0	1	0	11 User data bytes
				1	0	1	1	12 User data bytes
				1	1	0	0	13 User data bytes
				1	1	0	1	14 User data bytes
				1	1	1	0	15 User data bytes
				1	1	1	1	16 User data bytes
	1	1	1					Forwarding unlimited
	1	1	0					Forwarding up to 6 times
	1	0	1					Forwarding up to 5 times
	1	0	0					Forwarding up to 4 times
	0	1	1					Forwarding up to 3 times
	0	1	0					Forwarding up to 2 times
	0	0	1					Forwarding up to 1 time
	0	0	0					No more forwarding
0								Destination address is a physical address
1								Destination address is a group address

Table 3.35 The first two user data bytes for a switching command

Byte 6								Byte 7								Command
MSB							LSB	MSB							LSB	
D7	D6	D5	D4	D3	D2	D1 B9	D0 B8	D7 B7	D6 B6	D5 B5	D4 B4	D3 B3	D2 B2	D1 B1	D0 B0	
0	0	0	0	0	0	0	0	1	0	0	0	0	0	0	1	Switch on
0	0	0	0	0	0	0	0	1	0	0	0	0	0	0	0	Switch off

Table 3.36 The first two user data bytes for a switching command

Byte 6								Byte 7								Command
MSB							LSB	MSB							LSB	
D7	D6	D5	D4	D3	D2	D1 B9	D0 B8	D7 B7	D6 B6	D5 B5	D4 B4	D3 B3	D2 B2	D1 B1	D0 B0	
0	0	0	0	0	0	0	0	1	0	0	0	1	0	1	1	Dimming

Table 3.37 Control field

Byte 8								
D7	D6	D5	D4	D3	D2	D1	D0	
PB	PB	PB	PB	PB	PB	PB	PB	
0/1	0/1	0/1	0/1	0/1	0/1	0/1	0/1	Odd parity

Table 3.38 Acknowledgment Frame (with N1, N0: NACK bits; B1, B0: BUSY bits)

Byte 0								
D7 N1	D6 N0	D5	D4	D3 B1	D2 B0	D1	D0	
1	**1**	0	0	**1**	**1**	0	0	ACK
0	**0**	0	0	**1**	**1**	0	0	NACK
1	**1**	0	0	**0**	**0**	0	0	BUSY
0	**0**	0	0	**0**	**0**	0	0	NACK + BUSY

3.12 KNX Hardware

This section provides a basic and brief introduction to KNX hardware.

As discussed in Sect. 3.5, a KNX system consists of various bus devices, particularly system components (e.g., a 320 mA power supply), sensors (e.g., a four-gang switch sensor) and actuators (e.g., a six-gang switch actuator). The demands on the mechanical, electrical and electronic configuration of the devices vary greatly depending on where they are going to be used and what functions they are to carry out. As a result, there are a variety of hardware solutions available. For example, the device must have the appropriate transceiver to connect it to the transmission medium used (e.g., KNX.TP, KNX.PL, KNX.RF). The devices are also available in different designs, for example, flush, surface or rail-mounted.

KNX devices have both internal and external hardware. The external hardware comprises the housing and electrical connections. The internal hardware consists of various electronic components, of which the microcontroller (μC) is of particular importance. For detailed information on the external hardware, see the manufacturers' technical data sheets. See [8] for a detailed description of the internal hardware.

3.12.1 External Hardware

KNX devices are available in different designs: rail-mounted (RM), built-in (BI), flush-mounted (FM) and surface-mounted (SM) devices. The user selects the form that meets the installation's specification, in other words, how and where the device is to be installed (see Sect. 3.5.1).

Devices are also either compact or modular in design. A modular device comprises an application module attached to a bus coupling unit (see Sect. 3.5.1).

As well as the mechanical features (form, housing, measurements, etc.), the electrical features also play an important role (e.g., voltage supply, degree of protection). Electricians must strictly follow the circuit diagrams and connection specifications for each KNX device—see the manufacturer's technical data sheets. Particular attention should be paid to the electrical specifications, for example, the maximum possible power that an actuator can switch to an output channel.

Figure 3.30 shows the circuit diagram and Fig. 3.31 shows the corresponding connection specifications for a six-gang switch actuator.

As well as the connecting terminals for outputs A to F, the six-gang switch actuator also has a programming button, a programming LED, and a bus terminal for connecting the two-core bus cable. The maximum current switched at each output is 10 A/AC.

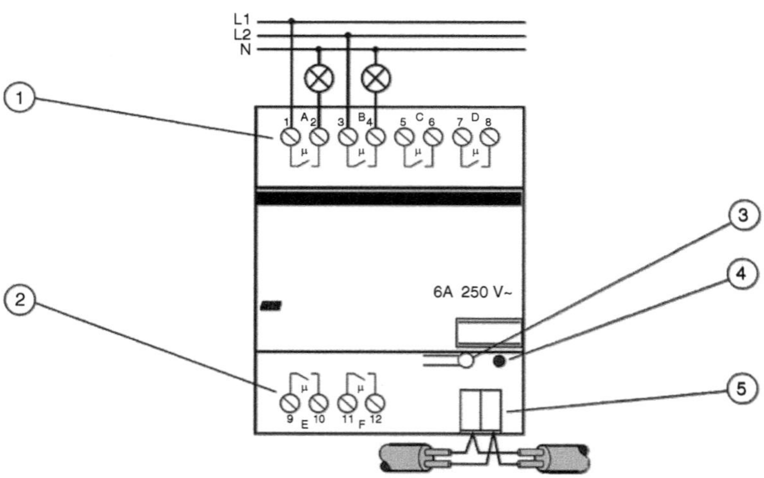

Fig. 3.30 Circuit diagram for a six-gang switch actuator [9]

Technical Data

Power supply	– EIB	24 VDC, via the bus line
Outputs	– 6 floating contacts	
	– Switching voltage	230 VAC +10/–15 %, 50 ... 60 Hz
	– Switching current	10 A/AC1
	– Basic time delay on single operation	typically 20 ms per relay
Operating and display elements	– red LED and push button	for assigning the physical address
Connections	– Load circuit	two screw terminals each, Wire range 0.5 ... 2.5 mm² finely-stranded 0.5 ... 4.0 mm² single-core
	– EIB	Bus terminal
Type of protection	– IP 20, EN 60 529	
Ambient temperature range	– Operation	- 5 °C ... 45 °C
	– Storage	- 25 °C ... 55 °C
	– Transport	- 25 °C ... 70 °C
Design	– modular installation device, pro*M*	
Housing, colour	– Plastic housing, grey	
Mounting	– on 35 mm mounting rail, DIN EN 50022	
Dimensions	– 90 x 72 x 64 mm (H x W x D)	
Mounting depth/width	– 68 mm / 4 modules at 18 mm	
Weight	– 0.24 kg	
Certification	– EIB-certified	
CE norm	– in accordance with the EMC guideline and the low voltage guideline	

Fig. 3.31 Technical data for a six-gang switch actuator [9]

3.12.2 *Internal Hardware*

You will only need a detailed understanding of a module's internal hardware, and the corresponding system and application software, if you develop KNX devices. For example, all devices that can communicate via the bus consist of two components: a communication module and an application module. The communication module contains a transmission module and a microcontroller (μC). Some sophisticated application modules also have microcontrollers. The configuration of the internal hardware must comply exactly with the KNX Association specifications otherwise it will not be certified [8].

A user does not need to know what the "internal parts" of a KNX device look like. Even so, the following section will describe a few essential features of the "internal" hardware, which will show you why KNX devices are intelligent communication devices. We will use KNX.TP in our example, because this is the most commonly used version of KNX.

3.12.2.1 The Principal Internal Configuration of a KNX.TP Communication Device

All devices that are to communicate via KNX.TP must comprise two main components (see Fig. 3.32):

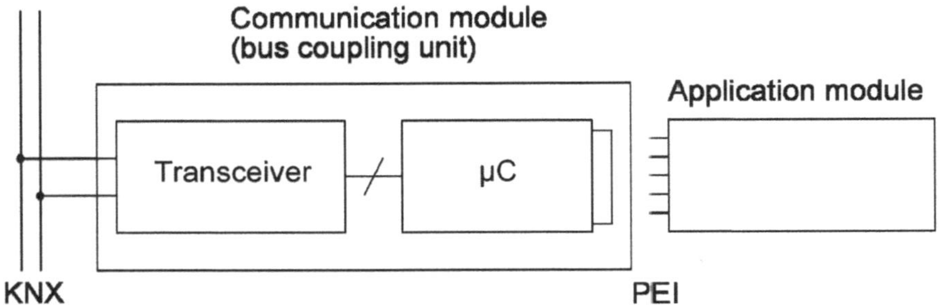

Fig. 3.32 Basic configuration of a KNX.TP application module

- A communication module
- An (integrated or modular) application module

The communication module consists of a transmission module (transceiver) and a microcontroller (µC).

If the communication module has a 10-pin (or 12-pin) physical external interface (PEI), it is also referred to as a bus coupling unit, see Sect. 3.12.2.4. A modular device comprises an application module that is plugged into the bus coupling unit. A compact device, on the other hand, comprises a communication module and an application module built into the same housing.

3.12.2.2 Transceivers

A device must always have a transceiver that is designed for the transfer medium that is to be used (KNX.TP, KNX.PL, KNX.RF, etc.). As standardized solutions, there are integrated circuits (IC) for (inductively) coupling a µC to KNX.TP. These solutions are certified and approved by the KNX Association, for example:

- Siemens (Infineon) FZE 1066
- Siemens EIB TP UART

The FZE 1066 contains analog electronics for sending and receiving data, reverse voltage protection, circuits for monitoring temperature and bus voltage, circuits for generating stabilized DC voltage (24 and 5 V) for both the µC and the application module, and a µC interface such as MC68HC05B6 and MC68HC705BE12 from Motorola.

The EIB TP UART enables easy access to the bus because it communicates with the host controller via a serial RS-232 interface. Figure 3.33 shows a simple example.

The application module consists of a simple switch that is opened or closed by, for example, a brightness sensor. When the contact is closed, the host controller (any µC with a RS-232 serial interface) sends a data frame to the TP UART,

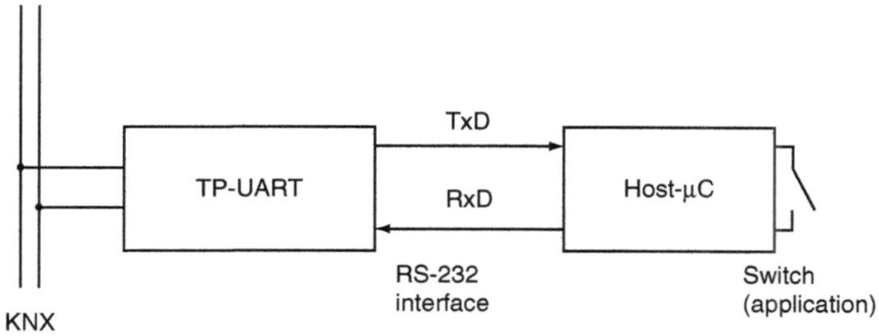

Fig. 3.33 Using a TP UART as a transceiver

observing the specified UART protocol. The TP UART then converts the data frame bits into a corresponding KNX.TP bus signal, which is then interpreted by, for example, a switch actuator.

A normal PC with Visual Basic could be used as a host controller. Pressing a button sends a data frame to the TP UART via the COM1 serial interface.

3.12.2.3 Microcontrollers

The type of μC used in a communication module always also depends on the functions that the KNX device is to carry out. The standard microcontrollers MC68HC05B6 and MC68HC705BE12—specially developed for KNX—vary depending on the size of ROM, RAM and EEPROM, for example. Both have ports for communicating with the application module. This interface is mechanically and electrically standardized (PEI see Sect. 3.12.2.4).

The KNX Association offers certified transceiver-μC components, such as the BIM M113 bus interface module [8] (see Fig. 3.34), to make it easy for developers

Fig. 3.34 Bus interface module BIM M113 [3]

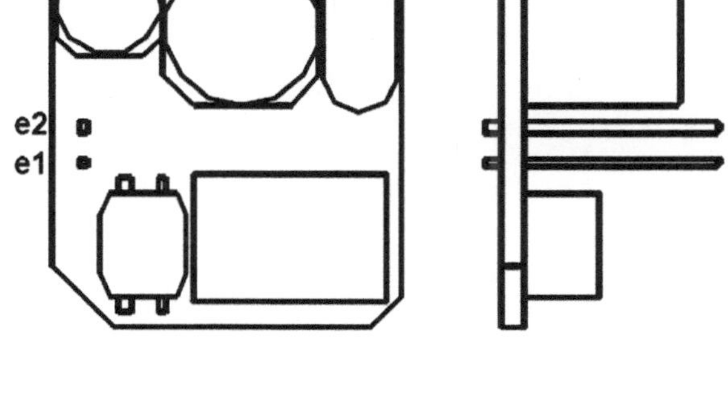

to have their KNX devices successfully certified and to also reduce development costs when they are implementing their own devices. The BIM M113 bus interface module has the FZE 1066, MC68HC705BE12 and system software (KNX BCU 2.1) on its circuit board.

3.12.2.4 The Physical External Interface

The electric and electronic design of an application module used for simple functions—such as executing requests from switch contacts or switching commands from relays—is very simple and consists of only a few components. The application software runs on the communication module's μC and the PEI is used as a simple input or output channel.

For more complex functions, the application software runs on the application module's μC and communicates with the communication module via the PEI's serial interface.

The PEI is a 10- or 12-pin connector (Fig. 3.35) and enables parallel and serial communication between the communication module and the application module.

Electronically, the pins have the following significance:

- Pin 1 and 10: Reference potential (0 V)
- Pin 6: Recognition of type (by using a type resistor in the application module)
- Pin 5: 5 V DC
- Pin 8: 24 V DC
- Pins 2, 3, 4, 7, 9: digital in-/outputs or connections for serial communication
- Pin 5a: pulse width modulation output
- Pin 6a: digital in-/output

For more details on the mechanical and electronical characteristics as well as communication protocols please refer to the standard [8].

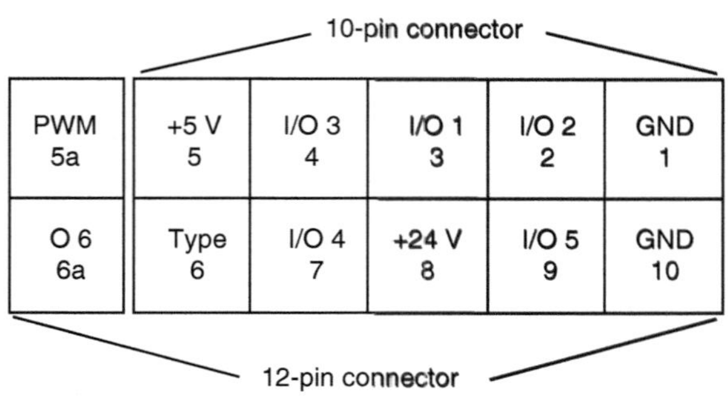

Fig. 3.35 Physical External Interface PEI (10- and 12-pin)

3.13 KNX Software

There are three types of software that can be used with KNX:

- System software
- Application software (application programs)
- Engineering Tool Software Version 5 (ETS 5)

This section will introduce you to the important aspects of KNX system and application software as well as the Engineering Tool Software Version 5 (ETS 5) program. KNX system and application software is used only by developers of KNX devices. Users, however, must be able to use ETS 5 and understand a few of the software concepts that arise when designing a project with ETS 5 such as setting device parameters and assigning communication objects to group addresses. For more detailed information, see [8] or [11].

3.13.1 Overview

KNX is a decentralized system. The system software and application software (also referred to as the application program) is incorporated into each KNX device.

- The system software, which is particularly responsible for processing data communication over the bus, is saved permanently in the device (in the communication module's ROM) by the manufacturer.
- A device's application software, on the other hand, is supplied by the manufacturer in an ETS 5 product database. The user must then select the corresponding product software, set the parameters, and then load it into the device using ETS 5 in order to commission the device.
- ETS 5 is an integrated program for designing and commissioning KNX systems.

KNX devices can usually run a number of applications. The user defines what function a device is to have by selecting the desired application software and setting the corresponding parameters, for example:

- By selecting the appropriate application software, you can set a coupler to function as line repeater (repeater application) or as a line or area coupler (coupler application).
- In the parameter dialog of a four-gang switch sensor, you can define whether a button is to work as an on/off switch, a dimmer or both.
- You can also set the parameters for a six-gang switch actuator so that an output is either normally open or normally closed.

To define the right application software and parameter settings for a KNX device, you first need to read the technical documentation closely. This can be extremely time consuming because a device may have large number of often complex applications and parameter options to choose from.

3.13.2 The Software Components in a Compact Device

Figure 3.36 shows the software components of a compact device that comprises a communication module and an application module in one unit. The device is supplied with the system software. As with the physical address, the application and its parameters must be programmed when commissioning the device.

The application software exchanges data with the system software using RAM flags and communication objects (C.Obj.). They send and receive data frames via the bus. RAM flags show the application software whether, for example, a data frame has been received and a communication object has been assigned a new value. The application can then access the C.Obj. and analyze its content. This communication also takes place in the same way but in the other direction. When the application has completed its task it accesses the parameters that the user set using ETS 5.

3.13.3 Software Components in a Modular Device

In a modular KNX device, comprising a bus coupling unit (as the communication module) and an (attachable) application module (AM), the application software, due to its complexity, is subdivided into an internal and an external application (see Fig. 3.37).

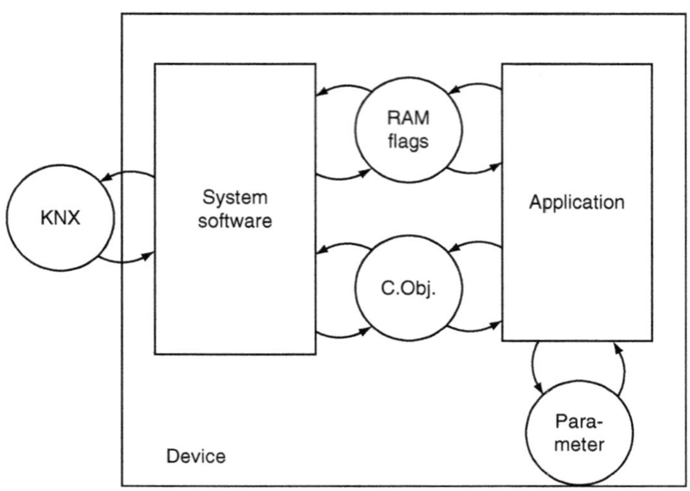

Fig. 3.36 Software components in a compact device

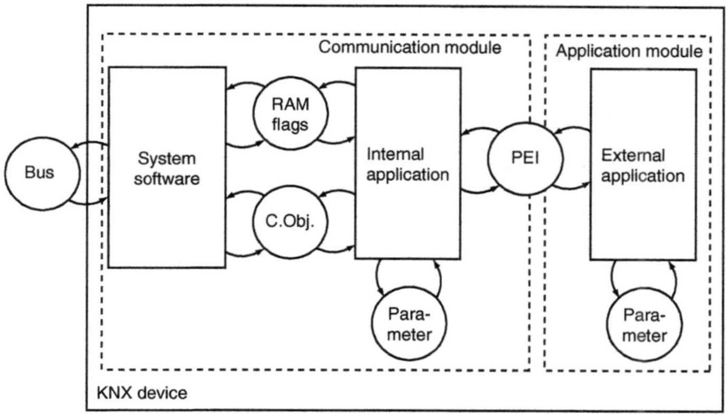

Fig. 3.37 Software components in a modular device

The external application (in the application module) is connected to the internal application (in the communication module) via the physical external interface (PEI). The internal application exchanges data with the system software using RAM flags and communication objects (C.Obj.). It reads and writes data frames from/on to the bus. The RAM flags show the internal application, for example, whether a data frame has been received and a C.Obj. has been assigned a new value. The internal application can then access the C.Obj. and transfer its content to the external application. This communication also occurs in the same way but in the other direction. As soon as the internal and external applications have completed their tasks they access the parameters set when the project was planned and designed.

When simple functions are implemented in the application module such as a request for rocker switch settings, then the information retrieval from the external sensors via the PEI can be fully processed as an internal application in the communication module.

With complicated functions (e.g., for a display), an external application runs on a separate microcontroller in the application module. The external application can then use the system software communication services in the communication module via the PEI Type 16 (serial synchronous interface). Sometimes you may need to download the external application and set its parameters using specific software from the manufacturer. In most cases, however, you can parameterize and program the application using ETS 5.

3.13.4 System Software

System software consists of a sequentially running component and an interrupt-driven component. Once the KNX device has been initialized by connecting it to the bus voltage, it carries out the following tasks:

- Launching the application
- Processing communication
- Starting check routines (monitoring the EEPROM)
- Managing the physical external interface

During initialization, the type of application module is defined and the device checks whether the programming button has been pressed and then whether the application program should not be launched.

The interrupt-driven component is responsible for processing the stream of bits received from the bus, and for initiating the SAVE routine should the power fail. The SAVE routine saves data in the EEPROM.

The user does not have to purchase the system software, because it is already saved to the KNX device's ROM on delivery. The KNX Association's certification ensures that devices from different manufacturers can communicate with each other in a KNX system. It also guarantees that the devices can be programmed using ETS 5.

3.13.5 Application Programs

3.13.5.1 Selection of Application

Application programs (or application for short) and their parameters determine a KNX device's functionality. A device has a number of applications that the user can choose from to carry out the desired building function(s). The manufacturer provides the application software in the form of a product database, which is available on CD or to download from the Internet.

Below are examples of applications for a four-gang switch sensor [9]:

- Switch LED
- Switch Dim LED
- Switch Shutter LED
- Switch Dim Shutter
- Switch Edge Flexible Allocation
- Value (EIS 6) LED

Depending on which application has been selected, a four-gang switch sensor sends either switching, dimming, shutter-control or 1-byte data frames.

As shown with the four-gang switch sensor, an application can execute a single function such as switching, or a number of functions simultaneously such as switching, dimming and adjusting the position of the shutters.

3.13.5.2 Application Parameters

Application programs have parameters that are set by the user using ETS 5 or specific plug-in software if the device is more complex. This process is called parameterization. The parameter dialog is often several pages long and can be very extensive and complex. The content varies depending on the type of device and the manufacturer. An application's parameter dialogs for a particular type of device can vary considerably in parts depending on the manufacturer! As a result, it is not always that easy to parameterize similar devices, such as switch actuators, from different manufactures in the same way, because the manufacturers have different operational and setting philosophies and different parameter options.

For example, after you have selected the *Switch Edge Flexible Allocation* application for a four-gang switch sensor, you can set one of the following responses to be executed when the upper push button is pressed [9]:

- No response
- Send a Switch ON data frame
- Send a Switch OFF data frame
- Send a toggle data frame

A further example of parameterization is defining LED colors for a switch sensor. An LED can show the value of the communication object that has been assigned to the button. For example, you can set the parameters so that the LED lights up red when the object value is 1 and green when the object value is 0. It can also be set to light up green and stay lit so that you can find it in the dark.

3.13.6 Engineering Tool Software, Version 5

Engineering Tool Software (ETS) is an integrated program for designing, planning and commissioning KNX systems. The design and usability of the software is similar to the ones known from Microsoft operating systems.

3.13.6.1 Licenses

ETS 5 can be downloaded from the konnex webpage (www.knx.org). The current version (as of June 2015) is version 5.0.6—build 1267. The software is always installed as a full version but has limited functionality depending on the license (see Table 3.39).

Table 3.39 Licenses for ETS 5 (as of October 2015)

License	Price	Functionality
Demo	Free	Maximum of 5 KNX devices per project, no export
Lite	200,00 EUR	Maximum of 20 KNX devices per project
Trainee	150,00 EUR	Only for students 1 project with a maximum of 20 devices, no export
Professional	100,000 EUR	Full functionality
Supplementary	100,00 EUR	Maximum of 2 licenses, only available for owners of a professional license

3.13.6.2 The Project Store

When installing ETS 5, a project store is automatically created at the location C:\ProgramData\KNX\ETS5\ProjectStore. This store is the project database or you can use it as the central database for all projects. It contains not only the product data from the manufacturer but also the project data supplied by you, the user. Before you select your devices and set their parameters, you must first import the product data into the project database from the manufacturer's product database (*Catalogs/Import…*), see Sect. 3.16 (Fig. 3.38).

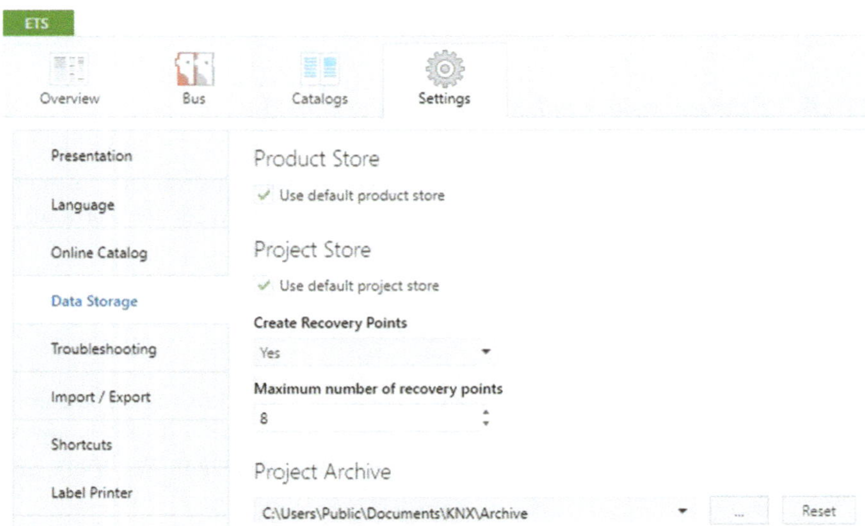

Fig. 3.38 Dialogue window for setting product data

3.13.6.3 Designing and Configuring Projects

After you have imported the product data, you can now design and configure your project using ETS 5. This involves converting the requirement specifications (customer's requirements) into functional (or target) specifications. In other words, determining the scope of supply and services, defining the layout and the types of devices, and creating the logical connections [12].

Designing and configuring projects with ETS 5 usually follows these steps:

- Create a new project. This opens the *building* window. Additional windows can be opened using the respective menu icon.

Now the project can be edited either in *building* view or in *topology* view:

- In the *topology* view the project can be seen as a bus topology. The following needs to be done:
  - Create areas and lines
  - Assign devices to lines. After assigning them their respective C.Obj. are visible
  - Assign physical addresses
  - Edit device parameters
- In the *building* view the project can be seen as a building structure. The following needs to be done:
  - Create buildings, building parts, rooms, control cabinets and systems
  - Assign devices to rooms, control cabinets or systems. After assigning them their respective C.Obj. are visible
  - Assign physical addresses
  - Edit device parameters

Afterwards switch to the *group addresses* window. The following needs to be done:

- Define group addresses to assign functions
- Assign communication objects to the group addresses using drag & drop. The assignment can be seen in the group address window and the topology window.

It is not mandatory to follow these steps in the given order. Steps can be left out or processed in another order.

3.13.6.4 Commissioning

After you have designed and configured the project, you will need to load the set physical addresses and the application program (device parameters, communication objects assigned to the group addresses and the actual application) into each device.

In order to do so, select the devices either in the topology window or the building window and start the desired process by right-clicking and selecting Download/…

When you program the devices for the first time, you must first load the physical address into each device (start the programming procedure in ETS 5 and then briefly push the device's programming button when prompted) and then load the application program for each device (including the device parameters and the assigned group addresses). You can also carry out both steps at the same time.

For the bus access to work, the used interface (serial PEI16-COM1, USB, if necessary IP gateway) has to be selected and configured in the tab interfaces of menu item ETS/Bus…

Afterwards the respective physical addresses have to be programmed. The selected interface is now displayed in the bottom left part of ETS 5 and can be edited there.

After the physical address has been programmed once, you only need to download the application software if the device parameters change or to select another application for a device.

Once you have loaded the application programs into the devices, the KNX system is then ready for use and can then be tested using ETS 5's diagnostic tools such as the bus monitor (menu item: *Bus/Bus Monitor*). This software lets you view and analyze the traffic on the bus. It records all the data frames sent on the bus.

3.14 Putting the Theory into Practice

To really get to know KNX, you should put all this theory into practice. For example, by constructing a small KNX system connected, via a USB interface, to a computer that will be used for designing and configuring the project. The next few sections will introduce you to such a system.

What projects is KNX mostly used for? This is an easy question to answer: KNX is mainly used for implementing lighting, shutter and blind control functions.

For those new to KNX technology, lighting control projects provide an excellent way of learning about KNX and how to use ETS 5. The knowledge gained can be relatively easily applied to cross-system projects such as heating functions. To learn more, you can go on product training courses (mostly free of charge) provided by the manufacturer and then use this knowledge to configure other devices and applications.

To implement small lighting control projects using KNX, you only need a few components. A basic KNX system needs only a power supply, an interface, a switch sensor, a switch actuator and, of course, the lamps we want to control. A safety extra-low voltage should be used for the load circuit, for example, ~24 VDC as the output of an isolation transformer.

The following system has been thoroughly tested while teaching KNX technology to electrical engineering and information technology students at the University of Applied Sciences in Mannheim, Germany (www.et.hs-mannheim.de).

Table 3.40 Devices for an EIB/KNX practice system

Manufacturer	Catalog reference	Type (ABB) or Article No. (BJE)
ABB	KNX Power supply, 320 mA, REG	SV/S 30.320.5
	Switch actuator, 10A, REG (6-gang)[a]	AT/S 6.6.1
	Bus coupling unit, FM	BA/U 3.2
	Switch (rocker) sensor, 4-gang, FM	BUTTON 4F, WS
BJE	REG Interface USB	6186 USB

[a]This switch actuator is no longer available. The ABB switch actuator SA/S 8.6.1 (8-gang), for example, can be used instead. For all current devices and prices, see the manufacturer's product catalog

We will use the KNX components from ABB and Busch-Jaeger Elektro (BJE) (Table 3.40).

To program and commission the system, you will need to purchase an ETS 5 license (with bus access) from the Konnex Association. For current prices for ETS 5 Professional (full version) or the full version for students, visit www.konnex.org.

To easily and safely transport the KNX basic system, the devices should each be housed in a solid casing (see Fig. 3.39). The KNX power supply is supplied with ~230 VAC via an IEC power socket. The 6 output channels A to F of the six-gang switch actuator can be connected using pin jacks.

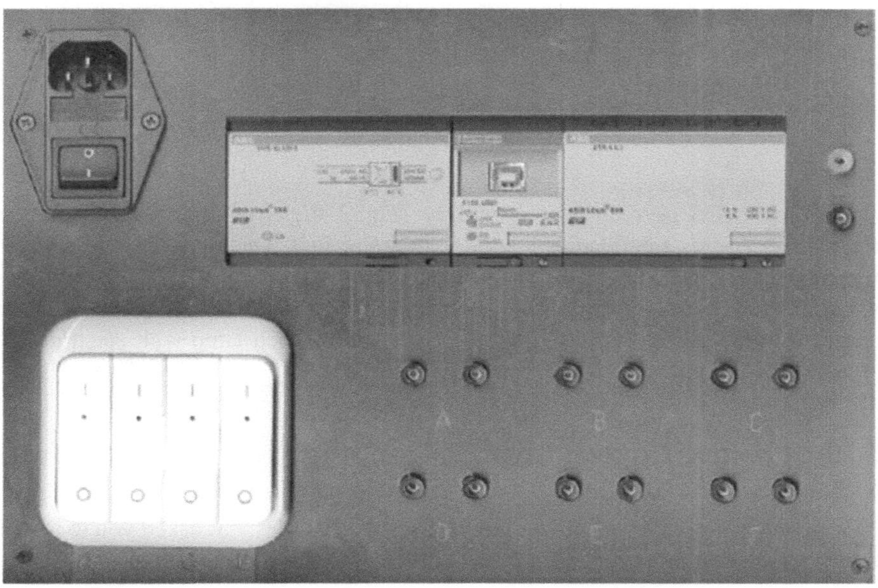

Fig. 3.39 A basic KNX system with an IEC power socket, power supply [9], USB interface [13], a six-gang switch actuator [9], a four-gang switch (rocker) sensor [13], and six output sockets (relays) A–F

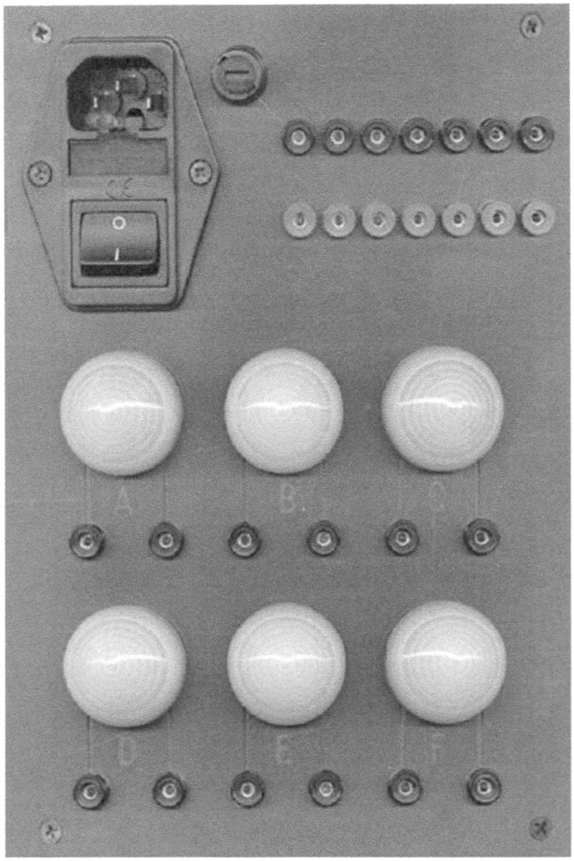

Fig. 3.40 A lamp assembly with an IEC power socket, seven output sockets with 24 VAC SELV and six lamps (A–F)

You will also need a lamp assembly (Fig. 3.40), comprising one isolation transformer and six small lamps housed in the same unit. The transformer's primary winding is supplied with ~ 230 VAC via an IEC power socket, and the secondary winding supplies a safety extra-low voltage (SELV) of ~24 VAV via seven black and seven blue power sockets. Two pin jacks supply each lamp with electricity.

3.15 Practice Project: Lighting Control

You are a KNX-certified electrician. A customer, Mr. Jones, has asked you to install a lighting control system in his house (Fig. 3.41).

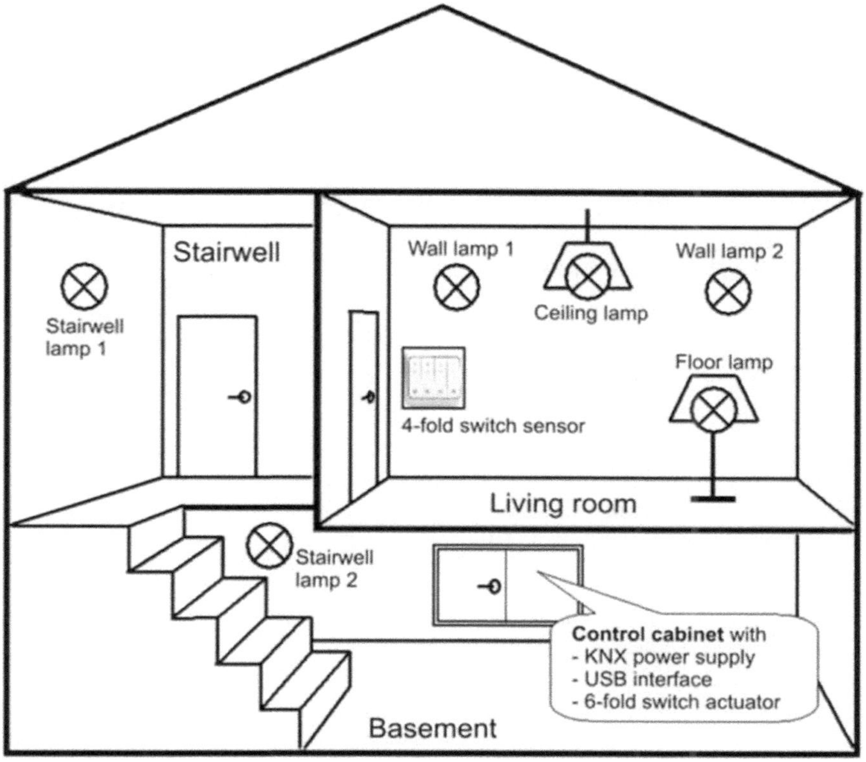

Fig. 3.41 A KNX lighting installation for a single-family house

3.15.1 Customer Order

Mr. Jones has ordered the following lighting functions (requirement specifications):

- One rocker is to switch the ceiling light (ceiling lamp) in the living room on or off. The upper contact should switch it on and the lower contact should switch it off.
- A second rocker should switch the wall lighting (wall lamp 1 and wall lamp 2) in the living room on or off. The upper contact should switch the lights on and the lower contact should switch the lights off.
- A third rocker should execute the following functions: Pressing the upper contact should switch the lamp on/off. Pressing the lower contact should switch on the lights in the stairwell. Stairwell lamp 1 should then turn off automatically after $\sim$2.5 s and stairwell lamp 2 after $\sim$5 s.
- Pressing the upper contact of a fourth rocker should switch on all the lights in the living room.

- Pressing the lower contact of this fourth rocker should switch off all the lights in the living room.
- Each rocker should have a green illuminating light.
- The rocker switches are to be installed in the living room. All the other KNX devices are to be installed in a control cabinet in the basement.

3.15.2 Required Devices

To be able to implement the lighting functions ordered by the customer using KNX, you will need the following KNX devices:

- Bus coupler, FM
- A four-gang flush-mounted switch sensor (clipped on the bus coupler)
- Cover frame (placed between bus coupler and switch sensor)
- Six-gang switch actuator, 10 A, REG
- KNX power supply, 320 mA, REG
- REG USB interface

(You could also, for example, use two two-gang switch sensor instead of one four-gang, or an eight-gang instead of the six-gang switch actuator, so that you could add load circuits (lamp circuits)).

You will also need a control cabinet, low-voltage and KNX cable and six lamps.

3.16 Designing and Configuring Projects Using ETS 5

3.16.1 Preliminary Considerations

You should always start by deciding which functions are to be assigned to which rocker switch on the four-gang switch, and which lamps are to be connected to which of the output channels on the six-gang switch actuator. Table 3.41 shows a potential solution.

How do we interpret the information in Table 3.41? Take the example on the first line, this states that pressing the upper contact of the left-hand rocker will send a *Switch ON* data frame to output channel A on the six-gang switch actuator, which in turn closes the ceiling lamp circuit, and the lamp lights up.

Figure 3.42 shows again which functions have been assigned to which rocker switch on the four-gang switch sensor. In the brackets are the output channels on the six-gang switch actuator that are used for the functions.

Table 3.41 Lighting control functions

Rockers on the four-gang switch sensor	Function	Output channel on the six-gang switch actuator	Function
Left upper	Switch on ceiling lamp	A	Close ceiling lamp circuit
Left lower	Switch off ceiling lamp	A	Open ceiling lamp circuit
Middle left upper	Switch on wall lamp 1	B	Close wall lamp 1 circuit
	Switch on wall lamp 2	C	Close wall lamp 2 circuit
Middle left lower	Switch off wall lamp 1	B	Open wall lamp 1 circuit
	Switch off wall lamp 2	C	Open wall lamp 2 circuit
Middle right upper	Switch floor lamp on/off	D	Close/open floor lamp circuit
Middle right lower	Switch on stairwell lamp 1	E	Close stairwell lamp 1 circuit[a]
	Switch on stairwell lamp 2	F	Close stairwell lamp 2 circuit[a]
Right upper	Switch on all living-room lamps	A–D	Close all living room circuits
Right lower	Switch off all living-room lamps	A–D	Open all living room circuits

[a]The actuator will automatically open the switch contacts for the output channels E and F after a specified amount of time (stairwell function)

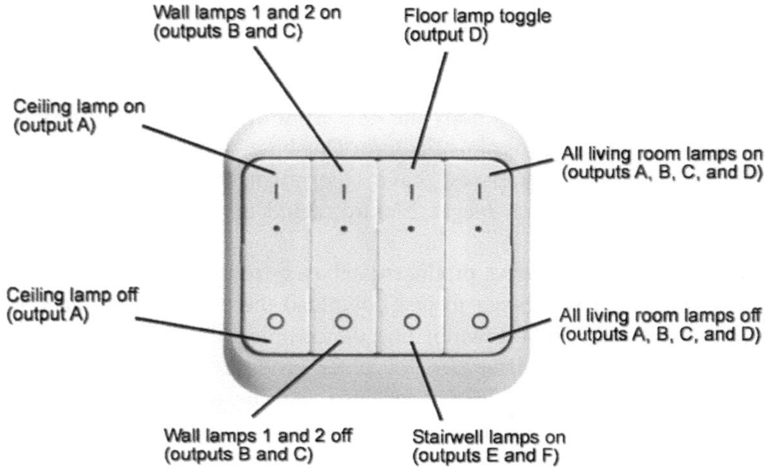

Fig. 3.42 The functions assigned to the rockers on the four-gang switch sensor along with the associated output channels on the six-gang switch actuator

3.16.2 Launching ETS 5

The ETS 5 start-up screen will open. The main work space is still empty and does not contain any windows. The next step is to create a new project.

3.16.3 Creating a New Project

A new project is created by selecting the green plus symbol (create new project) in the standard view. In the following dialog, a unique project name has to be defined e.g. "customer_1". After clicking *Create Project*, the ETS 5 workplace with the *building* window opens. By selecting menu item "Workplace/Open New Panel" additional views of project customer_1 can be opened:

- Topology
- Group addresses
- Project root
- Devices

First start with either the building window or the topology window. In the following section we will begin with the topology window.

3.16.4 Importing Product Data

In the topology window, you define the system's areas and lines, and add devices to the lines. Before you do this, however, you must first import the product data from the manufacturer's database for each of the devices to be used in the project. To do this, click on *Catalogs/Import...* A dialog box will open so that you can select the database of the manufacturer. You can import the data directly from the CD. We recommend, however, that you copy the data from the CD to a directory on your hard drive such as to C:\Programs\Ets3prof\Database, before you launch ETS 5. The ETS 5 product databases have the filename extension.VD3 (e.g., BJE_V32_0606.VD3) [Busch-Jaeger Elektro, dated 06/2006] or IBUS0505.VD3 ([9], dated 05/2005).

Always download the latest product databases from the manufacturers' Web sites. Sometimes it maybe easier to just download the databases for each product you intend to use, rather that downloading the complete database, because this can take quite a while. Some databases for older products only have the filename extensions.VD2 or even.VD1. ETS 5 can also read these files.

To find a product in the product database, you will need the product code. You will also need to determine which application programs you should use—see the

Table 3.42 The devices and product databases that are to be imported

Manufacturer	(C) Code, (A) Applications program	Individual product database
ABB	(C) SV/S30.320.5 power supply, 320 mA, REG	SV303205.VD1
	(A) -	
	(C) AT/S6.6.1 6f switch actuator, 10 A, REG	ATS661.VD1
	(A) Switch default stairwell function/3	
	(C) 4F, WS 4f switch sensor, FM	4F_WS.VD1
	(A) Switch slats flexible assignment/3.1	
BJE	(B) 6186 USB interface	ap_6186_USB_3_ger.vd3
	(A) USB interface/1	26.07.2006

product data sheet (specifications). This is can be found on the manufacturer's CD or can be downloaded from the manufacturer's Web site.

For our example we will need to import the four products and the application programs shown in Table 3.42. The power supply does not require an application program.

Mark the product in the *Import dialog*, for example, SV/S30.320.5 power supply, 320 mA, REG; and then click *Import* to import it into the project database.

3.16.5 Defining Areas and Lines and Adding Devices

After you have imported all the product data required for the project, you can then define the areas and lines and add devices to each line in the topology window.

For the customer_1 project, you can attach all the devices to just one line, which means you only need one area.

After you have marked the *Topology* item on the left of the topology screen, click *Edit/Add areas...* to add an area. Give the area a name (e.g., "House") and a number (e.g., "1").

Now mark the area *House 1* and then add a line to the area by clicking *Edit/Add lines...* Give the line a name (e.g., "Basement-1F") and a number (e.g., "1").

By clicking the + or − icons on the left-hand side of the screen, you can show or hide the areas and lines.

Now mark the line 1 *Basement-1F* and then click *Edit/Add devices* to open the *Catalog* window (see Fig. 3.43). You can use the product search function to view the devices in the project database. Now select, for example, the manufacturer and click *Search* to view all the products from this manufacturer.

Select the devices you require and click *Add* to add them to the line. The topology screen will now display the devices and application programs (see Fig. 3.44).

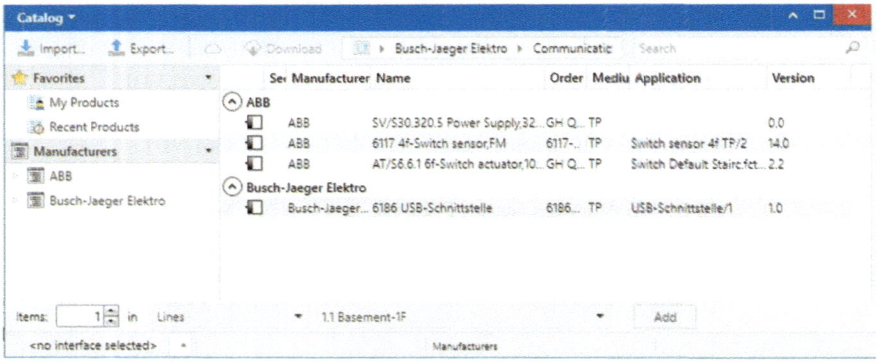

Fig. 3.43 ETS 5 product search dialog window

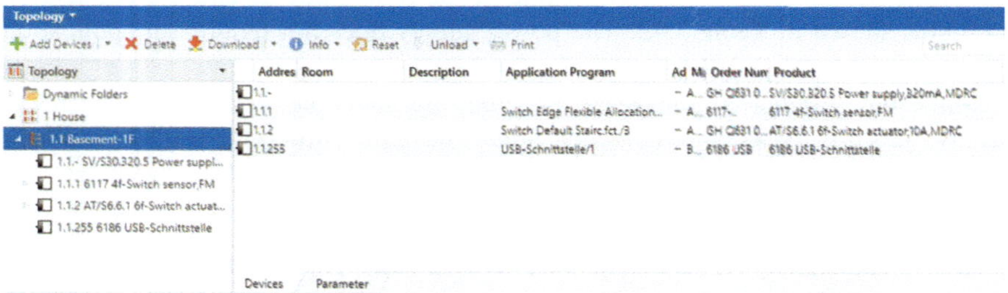

Fig. 3.44 The topology window of the *House* project showing all added devices

If a device has already been programmed and, therefore, already has a physical address, you can modify the physical addresses (here: 1.1.1, 1.1.2, and 1.1.3) that were automatically assigned by ETS 5 by marking each device and clicking *Edit/Properties...*

3.16.6 Setting Device Parameters

To ensure that the four-gang switch sensor and the six-gang switch actuator execute their functions correctly, you need to set their application program parameters. To open a device's parameter dialog, mark the device and select the *Parameter* Tab.

3.16.6.1 Parameter Dialog of a Four-Gang-Switch Sensor

Set the functions of the four LEDs to *Orientation light (green)* (Fig. 3.45). You now need to select the parameters for the push buttons. Set the parameters for the

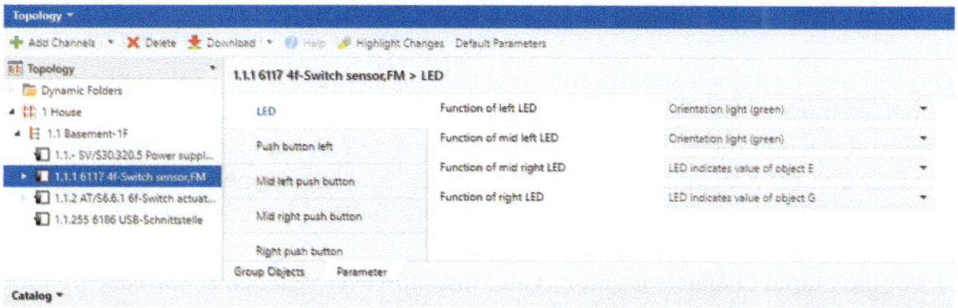

Fig. 3.45 The LED parameter dialog for the four-gang switch sensor

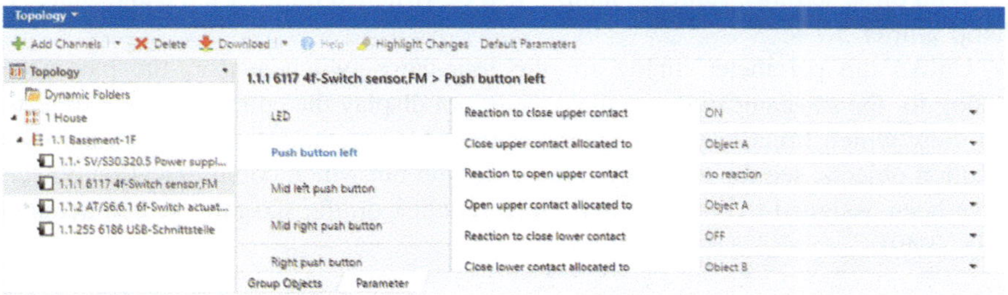

Fig. 3.46 Parameter dialog for the left push button of the four-gang switch sensor

left, *mid left* and *right* push buttons, as shown in Fig. 3.46 for the *left push button*: *reaction to close upper contact*: *ON*; *reaction to close lower contact*: *OFF*.

Set the parameters for the *mid right push button* as follows: *reaction to close upper contact*: *Toggle*, *reaction to close lower contact*: *Toggle*.

You can leave the default settings for the other push button parameters unchanged. Just make sure that you note which action is assigned to which communication object(s) (Table 3.43).

Note that for each rocker, you can also just use one object (e.g., *Object A* for the *left push button*) to transmit the switch information.

Table 3.43 Action and communication objects

Push button	"Action" connected to: communication object
Left	*Close upper contact* connected to: Object A
	Close lower contact connected to: Object B
Mid left	*Close upper contact* connected to: Object C
	Close lower contact connected to: Object D
Mid right	*Close upper contact* connected to: Object E
	Close lower contact connected to: Object F
Right	*Close upper contac*t connected to: Object G
	Close lower contact connected to: Object H

3.16.6.2 Parameter Dialog of a Six-Gang-Switch Actuator

You can use the default settings for output channels A to D (Fig. 3.47), because the switch attribute *normally opened contact* and the *normal operation* mode already comply the requirement specifications: a *Switch ON* command closes the contact and a *Switch OFF* command opens the contact. The requirement specifications do not state the preferred setting for a bus power failure; therefore use the *contact unchanged* default setting.

For the output channels E and F, you must, however, select the *staircase lighting function* operating mode (Fig. 3.48).

Table 3.44 shows the settings to switch off staircase lamp 1 (output E) after ~2.5 s and stairwell lamp 2 (output F) after ~5 s.

With these settings, staircase lamp 1 will be switched off after 2.6 s and staircase lamp 2 after 5.2 s.

Unlike the parameter dialog for the four-gang switch sensor, the parameter dialog for the six-gang switch actuator does not display the communication objects. To view which functions (e.g., switch output A) are connected to which communication objects, see the topology screen. To find out which communication objects have been assigned to which output channel, click on the six-gang switch actuator (Fig. 3.49).

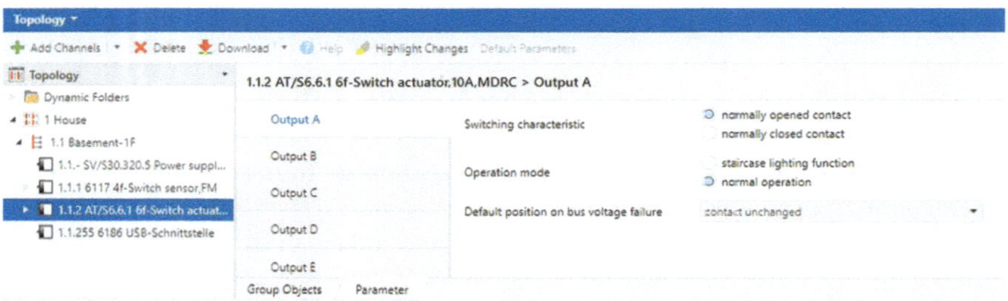

Fig. 3.47 The parameter dialog for output A of the six-gang switch actuator

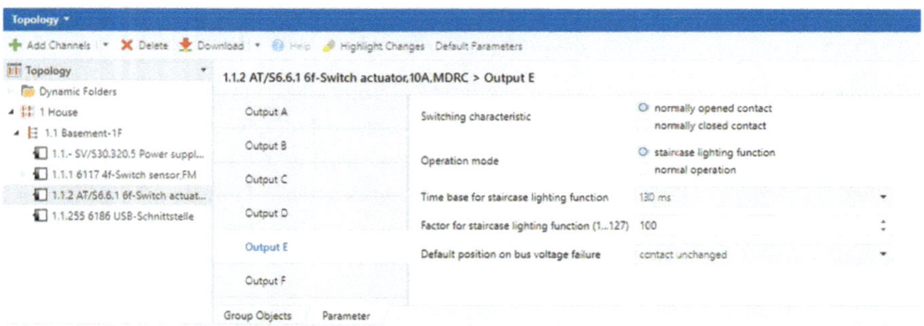

Fig. 3.48 The parameter dialog for output E for operation mode *staircase lightning function*

Table 3.44 Settings for the *staircase lighting function* operating mode

Output	Staircase lighting function	
	Time (ms)	Factor
E	130	20
F	130	40

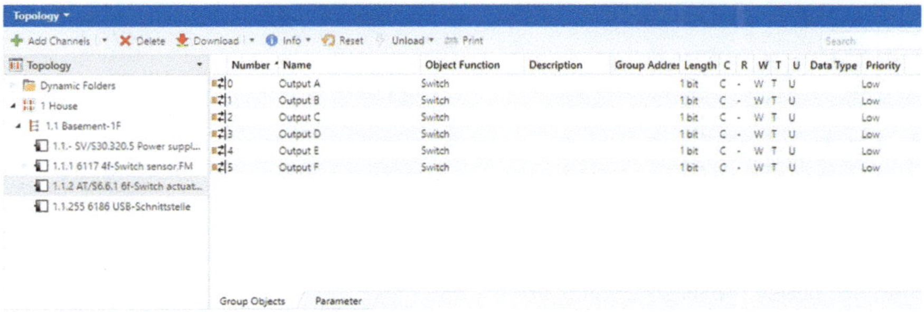

Fig. 3.49 The six-gang switch actuator's communication objects

Then assign the communication object with the number 0 (and the name *Output A*) to output A, the communication object with the number 1 (and the name *Output B*) to output B, and so on.

3.16.7 Creating Group Addresses

To connect the communication objects of the four-gang switch sensor with the communication objects of the six-gang switch actuator, you must first create group addresses for all the functions.

In the following, we will use two-level addressing (this setting can be changed when creating a project). For the desired functions (see specifications), we need to create the following group addresses (Table 3.45).

Table 3.45 Group addresses with main groups and subgroups

Group address	Main group	Subgroup
0/1	Central	All living room lamps on
0/2		All living room lamps off
1/1	Licht	Ceiling lamps living room on/off
1/2		Wall lamps living room on/off
1/3		Standing lamp living room toggle
1/4		Staircase lamps on

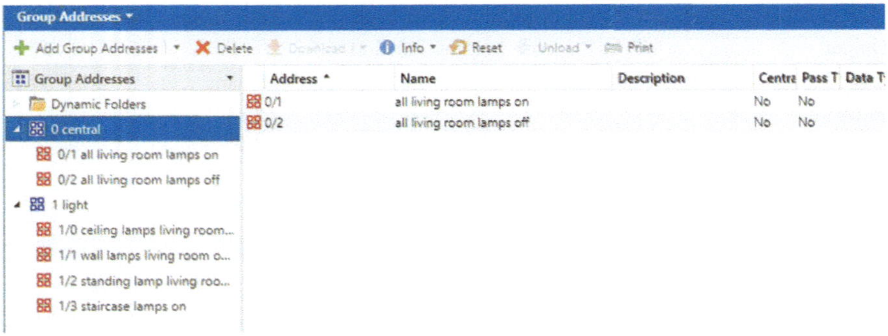

Fig. 3.50 Group addresses for customer_1 project

Using your mouse, mark *MainGroups* on the left-hand side of the group address window, then click *Edit/Add MainGroups* ... to add a main group. The most common names for these 2 main groups are:

- 0: central
- 1: light

To create a subgroup mark, for example, the *1 light* main group, and then click *Edit/Add GroupAddresses...* Give the subgroup a name that refers to the function to be carried out, for example, "ceiling lamp in the living room ON/OFF," and a number (address), preferably starting with "1." After you have created the group addresses, the group address window in the *customer_1* project should look like Fig. 3.50.

3.16.8 Assigning Communication Objects to Group Addresses

So that KNX can carry out the required functions, a communication object of at least one sensor must be connected with the corresponding communication object of at least one actuator via a group address.

To execute the *ceiling lamp in living room on/off* function, communication object 0 (object A—*switch data frame*) and communication object 1 (object B—*switch data frame*) of the four-gang switch sensor and communication object 0 (output A—*switch*) of the six-gang switch actuator must be assigned to group address 1/1.

The remaining communication objects are assigned to the other group addresses as shown in Table 3.46.

You can also assign communication objects to a group address by dragging and dropping. To do so you must make sure that the topology screen and the group

Table 3.46 Group addresses with assigned communication objects

Group addresses	(Sending) Communication objects for the four-gang switch sensor: Number (name)	(Receiving) Communication objects for the six-gang switch actuator: Number (name)
0/1	6 (object G)	0 (output A), 1 (output B), 2 (output C), 3 (output D)
0/2	7 (object h)	0 (output A), 1 (output B), 2 (output C), 3 (output D)
1/1	0 (object A), 1 (object B)	0 (output A)
1/2	2 (object C), 3 (object D)	1 (output B), 2 (output C)
1/3	4 (object E)	3 (output D)
1/4	5 (object F)	4 (output E), 5 (output F)

address screen are open and you can see communication objects belonging to the four-gang switch sensor and the six-gang switch actuator (Fig. 3.51).

After marking a group address, the C.Obj. can simply be dragged to the group adress window. The topology window also shows which group addresses the C. Obj. are assigned to when a device is selected. Figure 3.51 shows this state for the four-gang switch sensor after all assignments have been completed. Figure 3.52 shows all the group addresses (14) that the C.Obj. of the six-gang switch actuator are assigned to.

3.17 Commissioning

3.17.1 Hardware

To test the design and configuration of the lighting project, we will use the KNX basic system. You will be able to see that in order for the output channels A to F of the six-gang switch actuator to work correctly, the lamp assembly must be connected to the KNX assembly (this represents the electrically installed devices in a house). Each lamp has a special function (see Table 3.47).

Figure 3.53 shows how lamp 1 is connected to the safety extra-low voltage. The 24 V AC power supply is connected to output A of the six-gang switch actuator (relay as a switch), which is then connected to lamp A. The lamp in turn is connected to the power supply. We recommend that you use different colored cable, for example, black for the power supply to output, green to connect the output channel to the lamp, and blue to connect the lamp to the power source. Repeat the above for all the lamps and output channels.

After you have switched on the power supply to both assemblies and installed a USB cable to connect the KNX system's USB interface to the PC, you can now program and commission the four-gang switch sensor and the six-gang switch actuator using ETS 5.

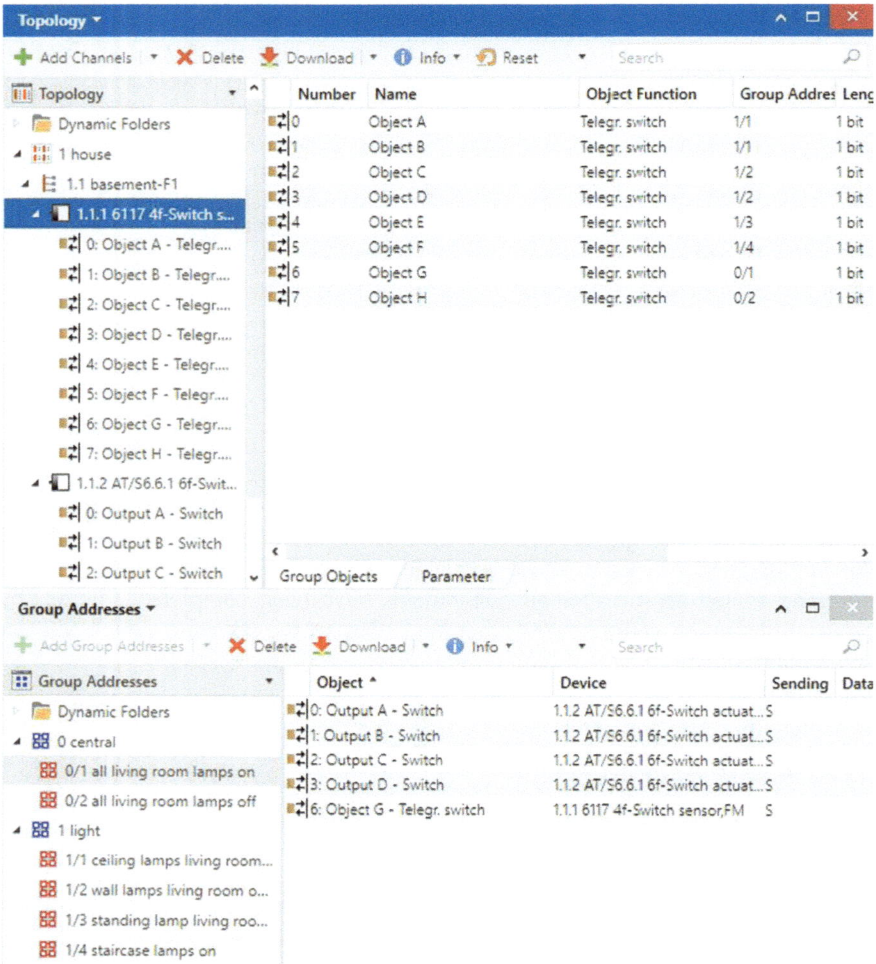

Fig. 3.51 Communication objects for the function “all living room lamps on”

3.17.2 *Programming Devices*

For the bus access to work correctly, the used interface (“serial PEI16-COM1” or “USB”) have to be selected, configured and (“locally”) programmed in the Bus tab of menu item ETS (see Sect. 3.13.6.4).

Mark a device in the topology window using your mouse, and select *Download*... to start the programming. When commissioning a device for the first time, you must first program its physical address. The access will happen via bus.

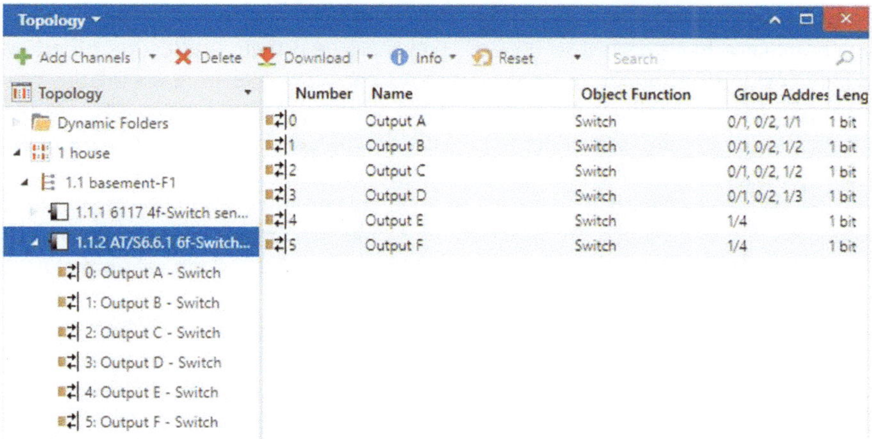

Fig. 3.52 Communication objects and group addresses for a six-gang switch actuator

Table 3.47 The functions of the lamps in test mode

Lamp	Function
1	Ceiling lamp
2	Wall lamp 1
3	Wall lamp 2
4	Floor lamp
5	Stairwell lamp 1
6	Stairwell lamp 2

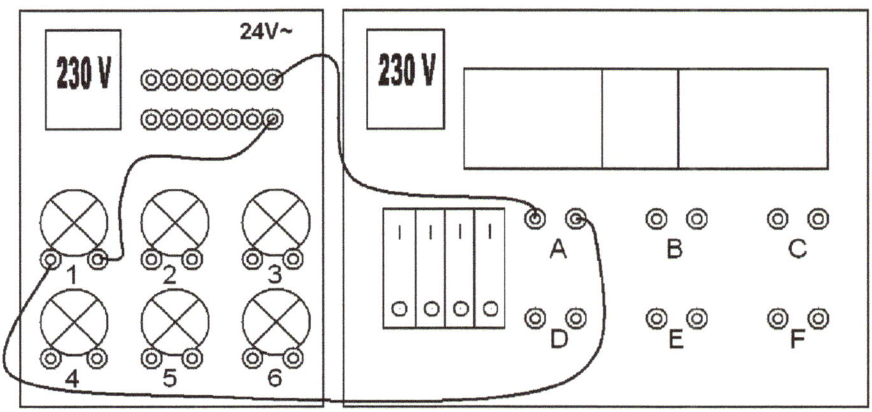

Fig. 3.53 Connecting the test lamps to the output channels of the six-gang switch actuator

After the physical addresses to the four-gang switch sensor and the six-gang switch actuator have been downloaded, you must load their application programs (click *Download Application Program*). You can view the progress of the programming process in the *Pending Operations* window (*Active* tab) (see Fig. 3.54). A typical programming procedure (e.g., fully downloading the application program for the six-gang switch actuator) takes ~15 s.

After you have changed parameters or assigned communication objects to group addresses, you can reduce the programming time by clicking ***Partial Download***. This reduces the overall programming time by only transferring the values that have actually been changed.

After you have downloaded the application programs, you can then test the system.

Fig. 3.54 The programme dialog

3.17.3 Testing the Lighting Control System

Once the system is ready for use, you test it to see whether it meets the requirement specifications. If one of the functions does not work properly, then you need to check or maybe reset the devices' parameters. You may also have to select another application program. All changes only become effective once the application programs have been downloaded to each device!

3.17.4 Diagnostics/Monitoring the Bus

ETS 5 comes with integrated software for monitoring and analyzing data frame traffic on the bus. To access this feature, click *Diagnostics* and then *BusMonitor*. By using special settings, you can view, for example, all data frames sent by a device with a specific source address. While the bus monitor is active, no devices can be programmed using the same interface.

Figure 3.55 lists the telegrams (data frames) that were sent by the four-gang switch sensor (source address 1.1.1) via the bus while the lighting control functions were being tested. As a group address always contains at least one of the six-gang switch actuator's communication objects, the actuator received, acknowledged and processed all the data frames.

The data frame with the number (#) 1 was sent when the upper contact of the *left push button* on the four-gang switch sensor was pressed. The data frame belongs to the group address 1/1, the routing counter is set to 6, the command is a *Write* command, and the data bit contains a "1" (in hexadecimal format: $01). The receiver—in this case the six-gang switch actuator—sent back a positive acknowledgment frame (LL-ACK).

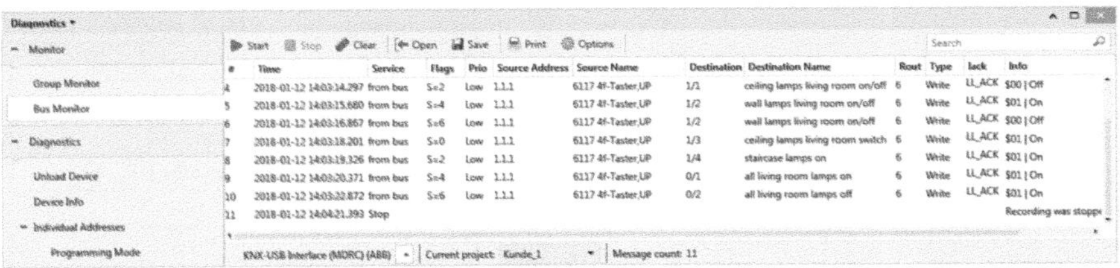

Fig. 3.55 Data frames shown while testing the lighting control function

In the column *flags* you can see information like e.g. S = 2 R P:

- S = 2 means, that 2 data frames where received (one data frame and one acknowledgment frame). For every frame received via the data interface, a counter starting at 0 is increased by 1. After reaching 7, the counter is reset to 0.
- R means, that the frame was a repeated frame.
- P means, that a transmission error occurred, which means, that at least one parity bit of the data frame was not identical to the receiver parity bit.

You can view the data frames in more detail by double-clicking on a particular line. Figure 3.56 shows more detailed information about the data frame (#) 1. The line with the name RawData seen in Fig. 3.56 contains the content of the data frame in hexadecimal format (Table 3.48).

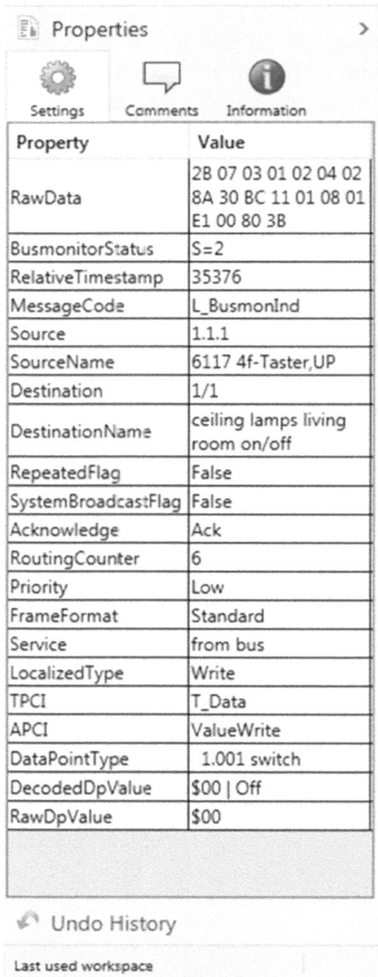

Fig. 3.56 Detailed information about data frame no. 1

Table 3.48 Content of the data frame

	Content of the data frame
2B 07 03 01 05 04 02 4F 94	BC 11 01 08 01 E1 00 81 3A

Table 3.49 Meaning of the data frame contents

Field	Hexadecimal code	Binary code	Meaning
Control field	BC	10 1 1 11 00	Low operation priority, no repetition
Source address	11 01	0001 0001 00000001	1.1.1
Destination address	08 01	0 0001 00000000001	1/1
DAF Routing counter Length	E1	1 110 0001	Destination address is a group address Routing counter: 6 Two user data bytes
User data	00 81	000000 0010 000001	Command „write“(0010) Real user data: one bit with value 1
Control field	3A	00111010	Control field

After translation to binary code, the meaning of the data frame contents can easily be identified (see Table 3.49).

3.18 Trends

KNX projects are becoming increasingly complex due to the larger number of sophisticated functions. Customers are expecting these installations to be easy to use. They want, for example:

- to be able to comfortably, conveniently and simply operate and monitor all functions from one or more locations
- to be able to modify the KNX installation themselves to meet their requirements, for example, changing the time the stairwell lighting remain on
- to have building control, multimedia, the Internet and telecommunication technology all connected.

3.18.1 Touch-Screen Control Panels

With design-oriented touch-screen displays, KNX manufacturers offer customers an increasingly broad spectrum of easy-to-use products that provide the desired functionality such as the control panel shown in Fig. 3.57. This control panel is also available as a comfort panel with even more functions.

This touch-screen control panel has the following features:

- It has up to 210 switch and control functions.
- You can connect any of the individual pages and configure them yourself.
- The screen displays up to 10 touch pads or eight large control buttons that you can activate using a PDA stylus or your finger tips. You can view, for example, the structure of the building or the functional groups.
- The integrated media pages when connected to an additional device enable you to control high-quality audio and video systems from the touch-screen control panel.

Touch-screen control panels provide you with excellent options that allow you to centrally control a KNX installation. Using the control panel shown in Fig. 3.58, you can switch on/off and dim all the lights on the first floor. The icon represents a lamp and by tapping the icon with your finger, you can switch the lamp on or off. If you press and hold your finger on the icon, you can dim the light. It also displays the status of the lamp, for example, if the icon is gray then the light is off and if it is yellow, then the light is on. You can also view, at a glance, all the status of the

Fig. 3.57 Touch-screen control panel [Busch-Jaeger Elektro]

Fig. 3.58 An image on a touch-screen control panel showing the first floor lighting control system in the CIBEK show house in Limburgerhof, Rhineland-Palatinate, Germany [14]

lighting on the first floor. Other pages enable you to control and monitor the lighting in the whole house.

3.18.2 Integrating Building Control into IP Networks

As well as a KNX installation, many buildings also have a local area network (LAN) connected to the Internet. You can connect these two systems and use the available network infrastructure, for example, to transfer data quickly between different areas. An IP Gateway is used to connect KNX systems to a LAN (see Fig. 3.59).

An IP Gateway is the interface between a KNX network and an IP network (see Chap. 5). It allows data to be exchanged between KNX devices and devices on the IP network. This enables you to, for example, integrate building control into building automation or remotely control a KNX system over the Internet.

An IP gateway can also be used as a line or an area coupler, enabling lines and areas to exchange data frames with each other over the LAN. This is particularly recommended when, for example, a control panel is connected to the area line, and it receives and reads a large amount of data from the KNX system. This can result in a considerable amount of data frame traffic (bottle neck) on the area line. A LAN transmits data far quicker (e.g., 10 Mbit/s) than KNX. Together with ETS 5 an IP

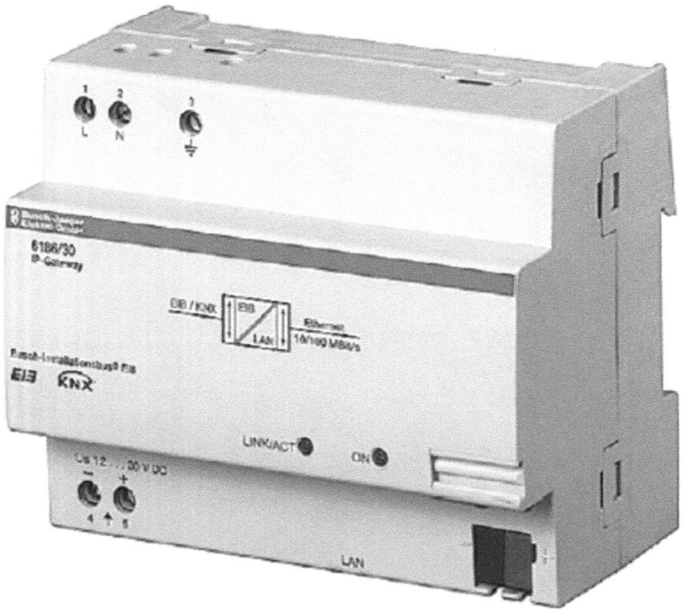

Fig. 3.59 IP Gateway [Busch-Jaeger Elektro]

gateway can be used to program KNX devices over the LAN, which means you no longer need serial or USB interfaces.

3.19 Exercises

Exercise 3.1
Plan a KNX system for the stairwell and corridor lighting in an apartment building as described in Sect. 3.2.2. The system should use 2-gang switch sensors and 6-gang switch actuators and include as few devices, sensors and actuators as possible.

Exercise 3.2
Name the different design forms for KNX devices.

Exercise 3.3
What is the purpose of a transceiver?

Exercise 3.4
What are modular KNX devices?

Exercise 3.5
Consider the crossover switching circuit shown in Fig. 3.60.
Is the lamp on or off?

Exercise 3.6
How many on/off, changeover and crossover switches are needed for the stairwell and corridor lighting described in Sect. 3.2.2, and what does the circuit diagram look like?

Exercise 3.7
What area are the bus devices in Fig. 3.22 assigned to? Where would you install a backbone coupler and what physical address would you assign to it, in order to couple the area to the backbone line?

Exercise 3.8
What is the maximum number of lines (lines, main lines, backbone line) a KNX system can have without having to use line repeaters?

Exercise 3.9
A KNX system comprises one line with 40 devices. The total cable length is 790 m. The greatest distance between a device and the power supply is 190 m. Does this system comply with the installation guidelines?

Exercise 3.10
How many sequences of UART characters does a data frame contain?

Exercise 3.11
How many pause bit times follow a UART character in a standard data frame?

Exercise 3.12
What is the bitrate for KNX.TP?

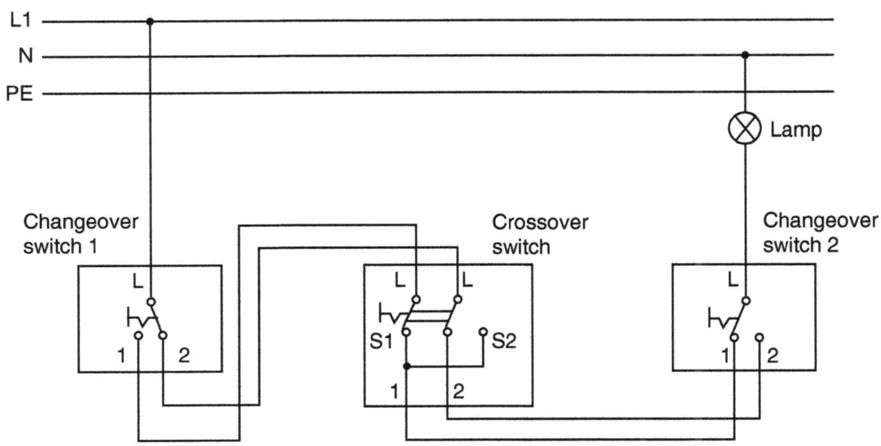

Fig. 3.60 A crossover switching circuit

Exercise 3.13
How much time does the transmission of a data frame with a length of 23 byte and access class 2 take?

Exercise 3.14
Which KNX device has the physical address 1.1.0?

Exercise 3.15
Can a KNX device have the physical address 10.0.48?

Exercise 3.16
A data frame has to be transmitted from sensor 1.1.23 to actuator 6.4.12. Which couplers (type, physical address) have to be used?

Exercise 3.17
How many UART characters does an acknowledgment frame consist of?

Exercise 3.18
Can an acknowledgment frame sent by two actuators on the bus contain both ACK and BUSY at the same time?

Exercise 3.19
Four receivers acknowledge receiving a data frame by sending a summation frame. Two receivers have received the data frame, one was BUSY and one received a transmission error (NAcK). Which acknowledgment frame does the sender detect (hexadecimal format)?

Exercise 3.20
A data frame (see Table 3.50) is stored using cross parity. What are the values for the data bits in the checksum field and the parity bits of the UART characters?

Table 3.50 Cross parity

Field	Data bits (D7 … D0)	Parity bit
Control field	10111100	?
Source address	00101010	?
Source address	00100110	?
Destination address	01001010	?
Destination address	00101110	?
DAF/routing/length	11100001	?
User data	00000000	?
User data	10001011	?
Checksum field	????????	?

References

1. Staub R, Kranz HR (2001) Raumautomation im Bürogebäude. Die Bibliothek der Technik, Band 210. moderne industrie, Landsberg/Lech
2. Kranz RUA (1997) Building control. Renningen-Malmsheim: expert
3. www.automation.siemens.com/et/gamma/>Support
4. Merz H (Hrsg) (2000) Kommunikationssysteme für die Gebäudeautomation – Grundlagen, Anwendungen, Projekte. Shaker, Aachen
5. Merz H (Hrsg) (2001) Kommunikationssysteme für die Gebäudeautomation – Theoretische Grundlagen und Praxisbeispiele. Shaker, Aachen
6. Merz H, Hansemann T (eds) (2003) Kommunikationssysteme für die Gebäudeautomation – Wirtschaftlicher Bedienungskomfort in Gebäuden mit Hilfe von Bussystemen. Aachen, Shaker
7. Rudolph W (1999) Einführung in die DIN VDE 0100. VDE, Berlin, Offenbach
8. KNX Handbook Version 2.0 Konnex Association (2009)
9. www.abb.de/knx
10. Rosch R, Dostert K, Lehmann K, Zapp R (1998) Gebäudesystemtechnik – Datenübertragung auf dem 230-V-Netz. moderne industrie, Landsberg/Lech
11. Dietrich D, Kastner W, Sauter T (2000) EIB Gebäudebussystem. Hüthig, Heidelberg
12. Handbuch Gebäudesystemtechnik. Frankfurt a. M.: ZVEI (Zentralverband Elektrotechnik- und Elektronikindustrie e.V., Fachverband Installationsgeräte und -systeme) und ZVEH (Zentralverband der Deutschen Elektrohandwerke) (1997)
13. BJE—Busch-Jaeger Elektro (2009) Technische Unterlagen zu KNX-Geräten. Lüdenscheid
14. www.cibek.de

Chapter 4
Building Automation with LonWorks®

The performance of the LonWorks® technology has already been pointed out in Sect. 1.7.2.2. It is an open networking solution for building automation and control networks that was developed by the American company Echelon. It is designed in such a way that it can be used in centralized building automation controllers as well as in decentralized building control components.

LonWorks is a standardized bus system (ANSI/CEA-709.1-B and EN ISO/IEC 14908) that enables intelligent devices to communicate with each other over a locally operated control network. LON stands for local operating network.

4.1 Technological Transition in Building Automation

Over the years building automation has been dominated by proprietary developments. An array of components and system applications were developed for monitoring, controlling and regulating building services such as ventilation and air-conditioning systems. This technology evolved from the components and system architecture used in process automation and was then adapted for building automation.

Originally, each sensor or actuator was wired directly to a central control computer system (Fig. 4.1). This was referred to as central control technology (see Sect. 1.7).

The development of more powerful microcomputers in the 1980s meant that control functions could now be processed "on site" in the systems themselves. These microcomputers had the same functionality as the programmable logic controllers (PLCs) used in factory automation and were tailored for use in building control systems. They were referred to as direct digital controllers (DDCs, see Sect. 1.2), because they had an in-built digital processor and were installed directly in the system. The DDCs' integrated communication interface meant that they could replace the large amount of conduit wiring that, up until then, had been

H. Merz et al., *Building Automation*, Signals and Communication Technology,
https://doi.org/10.1007/978-3-319-73223-7_4

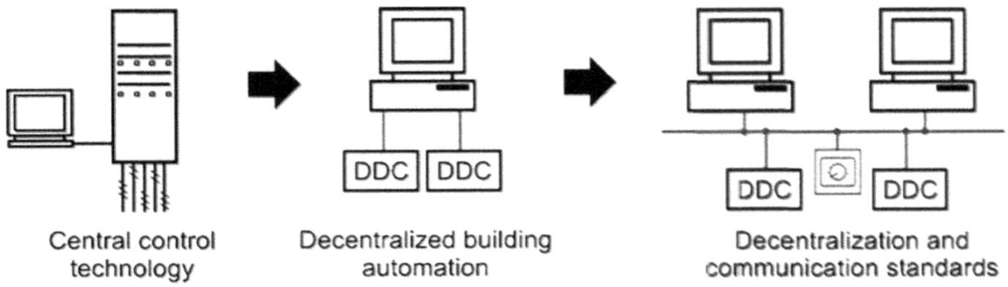

Fig. 4.1 Technological transition in building automation

necessary for connecting the various sensors and actuators to the control computer. These were referred to as decentralized solutions (Fig. 4.1).

A further step towards the decentralization of building automation was the advancements being made in semiconductor technology, resulting in smaller, more powerful microcomputers. They allowed even further decentralization of the functions for building automation. Sensors and actuators could now be equipped with local intelligence. A bus interface installed on the same circuit board made communication with other components possible. However, in order to provide this function for devices from different manufacturers, communication systems and protocols using open standards were required. Systems of this kind were first used for the control of lightning and blinds because they made the control more convenient and more flexible. The time for decentralized solutions and communication standards had arrived (Fig. 4.1). Soon, the standardized communication systems KNX (see Chap. 3) and LON gained a foothold in the German market. For the first time, you were not just confined to one manufacturer. You were free to choose the product that best suited your requirements from a wide range of products offered by various manufacturers, such as the room operating device with a LON interface shown in Fig. 4.2.

Nowadays both systems are well established in the market. In the area of building automation, LON and KNX are competitors. Both systems use a similar approach for the design of systems. However, LON and Konnex can also be used together, for example, mounting a KNX operating front on to a LON bus coupling unit (Fig. 4.3).

4.2 The Benefits of LonWorks® Technology

4.2.1 Use in Building Control

LON technology has become an integral part of building control. Distributing measurement, monitoring and control functions among local, decentrally installed components means that you can customize a variety of comfort solutions for an

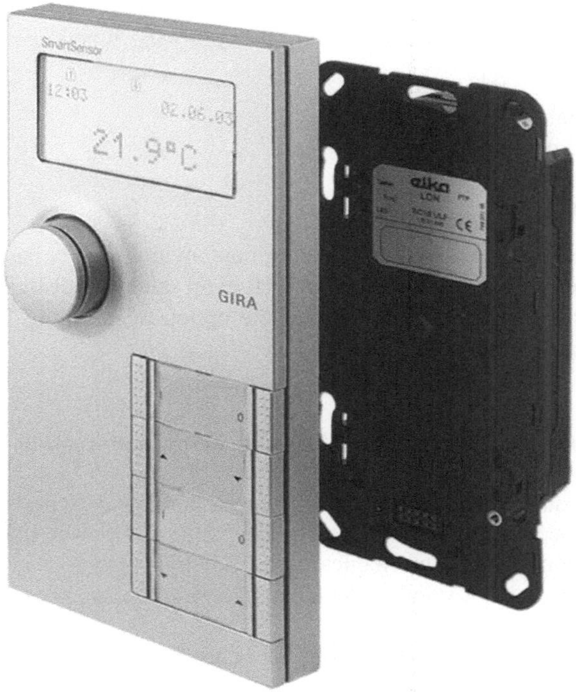

Fig. 4.2 Room operating device with LON interface for decentral installation [5, 6]

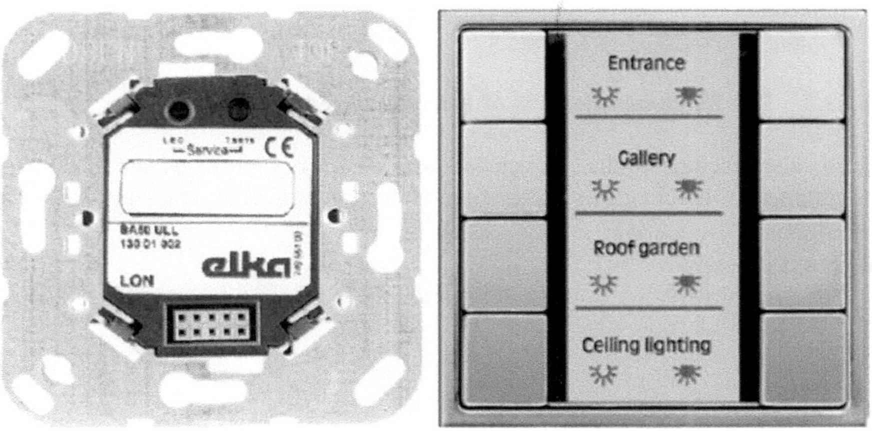

Fig. 4.3 LON bus coupling unit (left) and a KNX operation panel (right) [5, 8]

individual room. For instance, there is a wide range of sensors and actuators available on the market for implementing the following functions:

- Heating, cooling and ventilating
- Lighting control
- Shade/blind control
- Security
- Multimedia.

Building control focuses mainly on room automation and is becoming increasingly important in private residences as well as in commercial buildings. For this reason, some turnkey manufacturers offer the option of installing LON instead of conventional installation technology.

4.2.1.1 Replacing Conventional Wiring in Room Automation

Conventional control systems in individual rooms are always wire-based. Light and shutter control commands are sent over the traditional power line network. For instance, a wire connects a mechanical light switch directly to a lamp. As soon as the switch is turned on, it supplies the lamp with electrical energy (Fig. 4.4) and the light comes on.

Conventional installation technology is relatively simple to install and commission. But if you want to implement functions that run automatically, you will need a vast amount of additional wiring, not to mention the extra time spent installing the wires. Due to their complexity, comfort functions—for example, switching different lights in a room on or off from just one light switch or assigning different set commands to the lights' dimming components—are almost impossible to implement using conventional technology.

4.2.1.2 The Economic Benefits of Interconnected Systems

LON also facilitates the implementation of energy management functions. A commercial building is seen as a product in its own right and must therefore

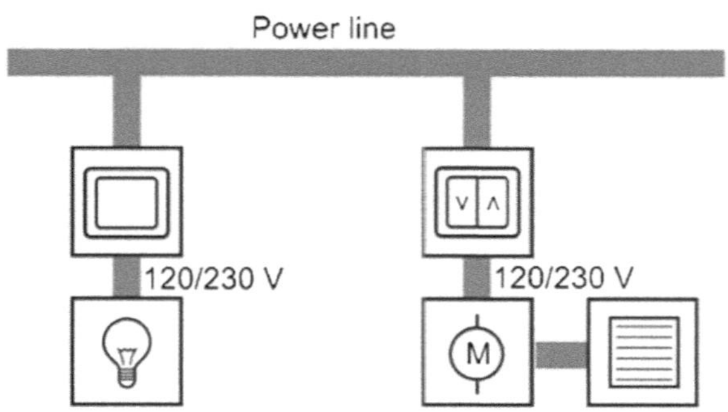

Fig. 4.4 Example of conventional installation technology

operate efficiently. If you want to set a timer-switch program in a conventional installation to switch off all the lights in an office building at the end of the day, then you would need to connect all the lights directly to the device running this program. This would require a vast amount of wiring. Furthermore, conventional installation technology does not allow you to connect all the systems so that you can reduce the setpoint temperature of the thermostats in all of the unoccupied rooms. This is the key advantage of the decentralized architecture of LON technology. It enables you to create a functional network by connecting all the systems (heating, ventilation, lighting, etc.) in a building [7].

As shown in Fig. 4.5, the principle of LON technology is that it separates the sensors and switches on one side from the actuators on the other side where the commands are executed.

Sensors and actuators communicate with each other by sending messages (data frames) over a bus. The power line cables are connected to the actuators on site, but do not come into contact with the sensors. This enables you to also control lighting and heating functions using a single device. Figure 4.6 shows a setpoint adjuster with an in-built temperature sensor and presence button.

Scenario: A person presses the presence button upon entering the room, illuminating the LED next to it. A command is then sent over the bus to activate the switch/dimming actuators next to the lamps and to start a constant lighting control program linked to a brightness sensor. This program is stored directly in the switch/dimming actuator. At the same time, the room's setpoint temperature is raised (e.g., to about 20 °C). The control program in the setpoint adjuster compares the current room temperature with this setpoint value. If there is a difference, it transmits a signal over the bus to the electronic actuator in the radiator telling it to increase or decrease the temperature accordingly.

You can also have the current status and values sent to the building automaton control computer. This information can be used by a superordinate timer-switch program in this main control computer to switch off all the lights and reduce the setpoint temperatures in all the rooms in a building at the end of the day.

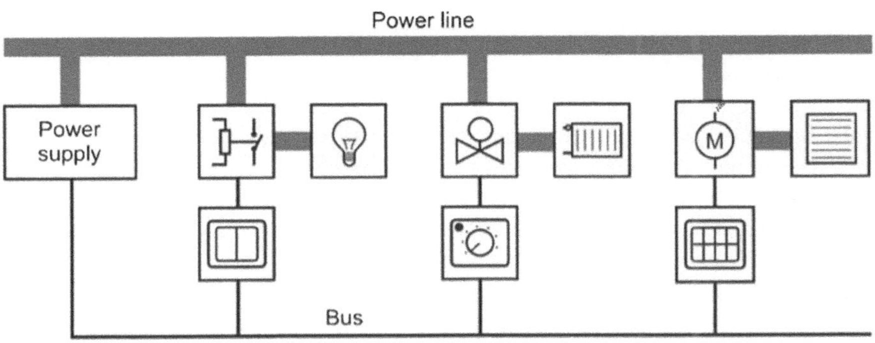

Fig. 4.5 The principle of LON technology in room automation

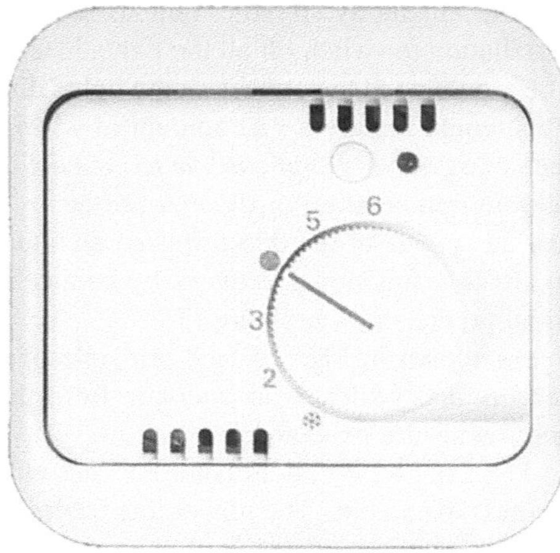

Fig. 4.6 A setpoint adjuster with a built-in temperature sensor and presence button [1, 13]

4.2.1.3 Increased Flexibility Through Reprogramming Instead of Rewiring

At some point you may need to use a particular room or building for a different purpose. You may want to convert, for instance, a conference room in a commercial building into two smaller offices or a study in your house into a children's bedroom.

In the conference room, instead of one light switch for all the lights, you now need two light switches for two separate circuits (one for each of the new offices). In a conventional installation, you would have to rewire the room to create the two separate circuits. In building control, however, LON technology enables you to make the required changes by simply reprogramming instead of rewiring, which usually far quicker. Furthermore, the LON programming tool also provides you with the new documentation.

4.2.1.4 Additional Security Features

By allowing you to connect components from the various systems or to send a single command to a number of bus devices, LON offers additional security options that are above and beyond the scope of conventional installations. For example, the signal from a smoke detector in an office room can be coupled with the actuators in the electrical windows, and doors and ventilation system. In the event of a fire, you can therefore ensure that office doors are automatically closed, the ventilation system is switched off and the windows are opened to extract the smoke. Figure 4.7 shows a LON device used for controlling the airflow in a ventilation unit.

With LON technology you can logically connect all the switch/dimming actuators in a house to one switch, creating a panic button. If you then hear an intruder

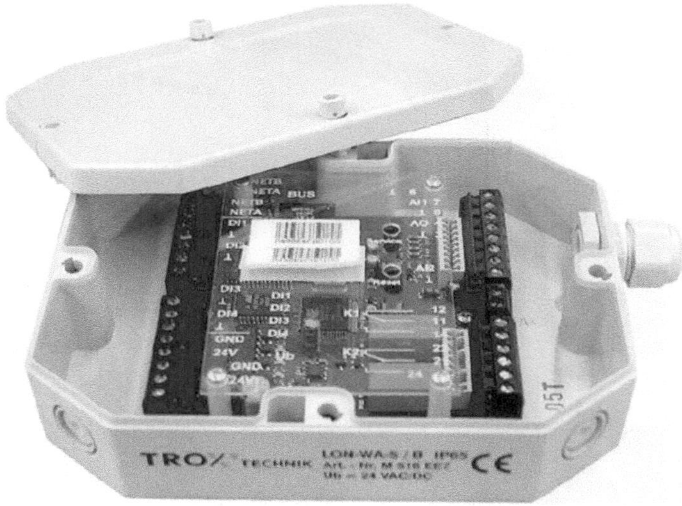

Fig. 4.7 LON component for controlling an airflow regulator [15]

at night, you simply flick this switch to turn on all the lights in the house. You could also logically connect this switch to your burglar alarm.

In summary, using LON technology in building control provides the following benefits:

- Comfort and convenience
- Efficiency through energy management functions
- Flexibility through the reprogramming options available
- Security.

4.2.2 Using LON Technology at the Automation Level

In building control, LON technology is used primarily for the decentralized processing of automation functions in room automation. You can use different processors to scale the system's performance and capability, which means that LON can also be used at the automation level. This level can carry out monitoring, controlling and regulating functions for building services such as heating and ventilation systems.

For example, if all the valve actuators for the static radiators in the offices have been turned up to their highest setting (fully open), this information is sent to the central heating system's DDC (Fig. 4.8), which then raises the flow temperature of the hot water to meet the increase in demand.

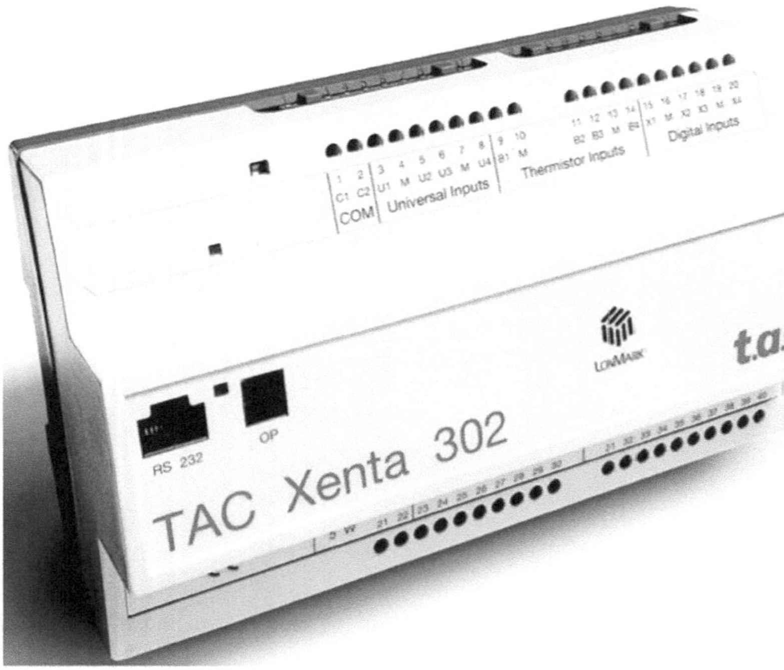

Fig. 4.8 A DDC for a LON-based central heating system [12]

The focus of using LON technology at the automation level is not to decentralize individual functions, but to provide a standardized integrated bus system.

4.3 The History of LonWorks®

Mike Markkula, the co-founder of the computer company Apple, laid the foundations for LON technology in the 1980s. His idea was to build a microcomputer that was so small and inexpensive that it could be installed in household appliances such as coffee machines. It would also have an interface to allow it to communicate with other devices.

Together with Dr. Ken Oshmann, Mike Markkula founded Echelon in 1986, based in Palo Alto, California. The company's aim was to develop an open system for decentralized networks—the result was the LonWorks technology or LON for short. The key to LON is a microcontroller, the Neuron Chip, which was introduced in 1990 and initially produced by Toshiba and Motorola. Later, Cypress took over from Motorola as the second supplier.

This was followed in 1997 with the introduction of the LonWorks Network Services, a development and application platform for open applications and integration tools. This along with the introduction of a multivendor instance for defining standardized variables and applications led to the technology's breakthrough.

Today, Echelon has more than 300 employees worldwide-mainly in North America, Europe, China and Japan—who develop and sell tools, devices and customized applications [4].

4.3.1 The Use of LonWorks Technology Worldwide

Echolon's decision to make all its technology openly available resulted in the variety and number of applications and devices on the market. Vendors and users from all sectors on every continent have access to Neuron Chips and the development, programming and integration tools (Fig. 4.9).

Internationally, LON technology is predominantly used for industrial (35%) and building automation in commercial buildings (35%), home automation (15%), and transportation and public utility control networks (energy distribution) (15%). Echelon's product database contains more than 1300 products from various manufacturers in the aforementioned sectors.

4.3.2 Organizational Units

Alongside Echelon as the main manufacturer, different manufacturers and users of LON formed the LON user organization known as LonMark Germany today. This organization includes different workgroups on building automation, instruction and training, industry, cooling technology and system integration.

The internationally operating LonMark Interoperability Association [www.lonmark.com] assures the further improvement of the technologies and that these improvements are not limited to one specific manufacturer. More than 300

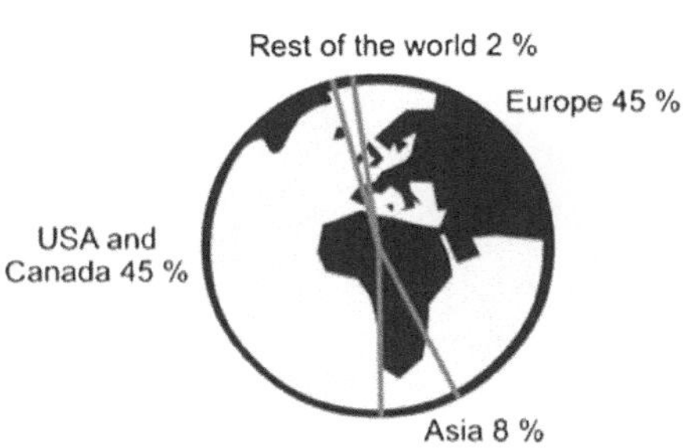

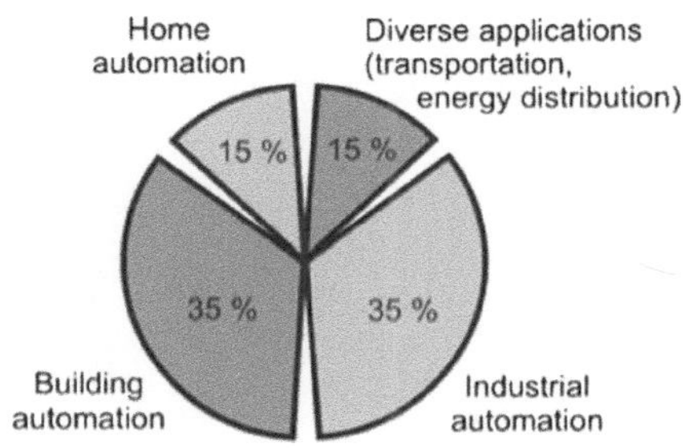

Fig. 4.9 LonWorks® technology worldwide

companies are organized in this association and its main task is to set standards for LON-variables and to define the requirements for newly developed components and parts.

4.3.3 Standardization

LonWorks has been published under the ANSI/CEA-709.1 LonWorks networking protocol specification, the ANSI/EIA-852, and the EN ISO/IEC 14908 series of standards.

4.4 Basics of the LonWorks System

4.4.1 Components

LonWorks technology consists of a vast number of interconnected elements, comprising hardware components, software applications and organizational structures (Fig. 4.10).

4.4.1.1 The Neuron Chip

The Neuron Chip is the heart of the LonWorks system (Fig. 4.11). It was developed by Echelon and comprises three processors that provide both communication and application processing capabilities. The Neuron Chip is one of the components

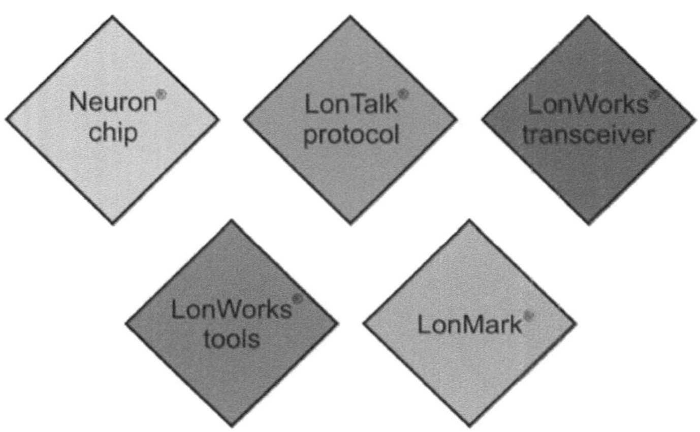

Fig. 4.10 Elements of the LonWorks technology

Fig. 4.11 The neuron chip [4]

of a LonWorks device. The name "neuron" was chosen to highlight the similarities between the topology of a LonWorks system and that of the neurons in a human brain. A Neuron Chip represents a node in a network.

4.4.1.2 LonTalk Protocol

The LonTalk protocol defines how Neuron Chips are programmed for different applications and how they communicate with each other as nodes in a network. This requires a standard language, or communication protocol. The LonTalk protocol is an integral part of the Neuron Chip and is embedded directly in the chip as firmware. This ensures that all LON nodes connected with each other on the same network are compatible.

4.4.1.3 Transceivers

Each network device has a transceiver—short for transmitter and receiver (Fig. 4.12). The transceiver provides a physical communication interface between a LonWorks device and a LonWorks network. There are specific transceivers for each medium: twisted pair, power line, radio frequency and fiber-optics. The most commonly used medium is twisted pair.

Fig. 4.12 A transceiver for connecting a node to the physical network [4]

4.4.1.4 LonWorks Tools

Echelon and other manufacturers offer various programming and integration tools for Lon technology. Firstly, Echelon provides development tools, LonBuilder and NodeBuilder, which allow developers to program their own applications into the Neuron Chips. These tools are used to develop and test LON devices. The average user can purchase ready-to-use LON devices with pre-installed applications and will not need to use these tools. Secondly, there are a variety of network tools available, such as LonMaker from Echelon, for customizing LON devices and integrating them into a fully functional network.

4.4.1.5 The LonMark Interoperability Association

Interoperability refers to the ability of devices from different manufacturers to be able to exchange information with each other. Ensuring the interoperability of components is extremely important. Lon technology enables the development of interoperable devices and systems. Manufacturers have their own ideas as to which functions devices should have so that they will sell. The LonMark Interoperability Association is responsible for defining a device's basic functionality and minimum requirements as well as the standard network variable types (SNVT). A device that has been certified by the LonMark Interoperability Association is guaranteed to be interoperable with other LonMark-certified devices from other manufacturers. The requirements are discussed in more detail in Sect. 4.5.5.

4.4.2 Components and Functionality of a LON Device

All LON devices possess the same basic components (see Fig. 4.14). We will now explain the importance of each of the components using the example of the setpoint adjuster with an in-built temperature sensor and presence button shown in Fig. 4.13 [2].

4.4.2.1 How a Neuron Chip (with Memory) Works

The Neuron Chip (Fig. 4.11) provides the LON device with its intelligence. It contains the LonTalk protocol. Its memory stores all programs that provide the functionality and run the desired applications. The types of Neuron Chip most commonly used are 3120 and 3150—made by Toshiba and Cypress. The Neuron Chip type 3120 is usually used in simple devices that do not carry out complex control functions.

The setpoint adjuster shown in interprets the signal from its in-built temperature sensor and transmits this signal as a variable over the bus. It does the same when you adjust the setpoint dial to a preset value or press the presence button.

The Neuron Chip 3150 is used in more sophisticated devices, such as the DDC in Fig. 4.9, that require more complex application programs. These devices are also fitted with additional external memory that can be used to store more complex control algorithms.

Fig. 4.13 A setpoint adjuster with a bus coupling unit [5]

Both types of Neuron Chips have three internal processors that each carry out different functions (Fig. 4.15).

- CPU 1 is responsible for physically accessing the transmission medium. A network interface provides access to the transceiver. This represents layers one (physical layer) and two (data link layer) of the ISO/OSI model (see Chap. 2).
- CPU 2 is responsible for transmitting network variables and represents layers three to six of the ISO/OSI model.
- CPU 3 processes application programs, but does not access the network; this is done by the other two CPUs. The CPUs exchange data with each other by accessing the shared random access memory (RAM).

The application programs and the configuration parameters selected during commissioning are stored in the electrically erasable programmable read-only memory (EEPROM). An example of configuration data is the preset room temperature setpoint in the setpoint adjuster shown in Fig. 4.13. The data in the EEPROM is not erased if the power is cut.

The read-only memory of a Neuron Chip 3120 stores the LonTalk protocol, the Neuron operating system and the predefined operating routines for input/output (I/O) conditioning. The Neuron Chip 3150 does not have internal ROM, but instead

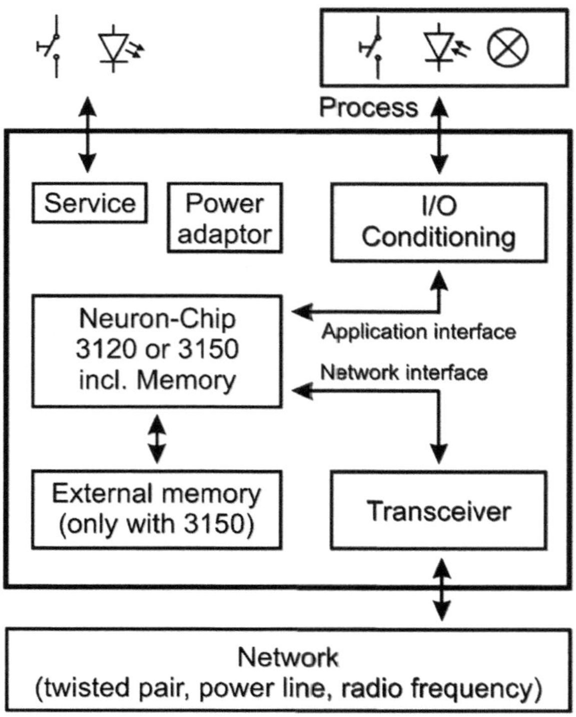

Fig. 4.14 The components of a LON device

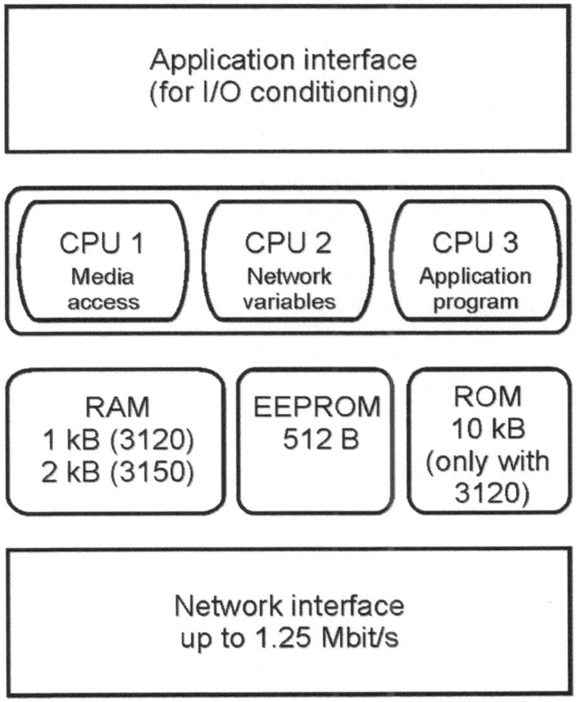

Fig. 4.15 Internal components of the 3120 and 3150 Neuron Chips

has external memory that stores the aforementioned EEPROM functions and still has enough storage space for running sophisticated application programs.

4.4.2.2 Input and Output Conditioning

A Neuron Chip needs special components to be able to communicate with other devices. For the processor to receive the signal from the temperature sensor on the circuit board, the signal must pass through an input unit (Fig. 4.14). An analog-to-digital converter then converts the sensor's temperature-dependant resistance value into an input byte value that the Neuron Chip can understand—this is called input/output (I/O) conditioning and varies depending on the manufacturer and the required application.

In the example of the setpoint adjuster in Fig. 4.13, the value from the potentiometer is recorded as well as the temperature value, and is then transferred to the processor as an input byte. If the device's presence button has been pressed, the I/O unit sends a one-bit signal to the processor. The processor then interprets this signal, transfers it on the bus as a variable, and then emits a one-bit signal over the output unit to activate the LED next to the button.

The application interface connects the Neuron Chip with the input and output units. The manufacturer can freely configured and customize the application interface to the device's functionality using preprogrammed operating system routines. The following functions are provided for configuration:

- Bit input and bit output
- Byte input and byte output
- Serial input and serial output
- Slope input
- Period and pulse count input
- Frequency output

Standardized interface control programs make it easier for developers to program Neuron Chips.

4.4.2.3 Power Supply and Adaptor

There are different ways of supplying a device with power. More complex devices that require a large amount of electrical power, such as the DDC in Fig. 4.8, have separate contacts over which an external voltage source supplies it with a DC voltage of 24 V. An easier option is to supply the device with power directly from the bus using a special transceiver. This is particularly suitable for building control components that do not require a large amount of power such as a setpoint adjuster. Another option is to connect the device directly to the main power line. The power line not only supplies the device with power (voltage) but can also be used to transfer data. This option also requires a special transceiver. In all examples, a power adaptor (Fig. 4.14) on the device's circuit board converts and stabilizes the required voltage.

4.4.2.4 The Service Button and Neuron ID

Pressing the device's service button (Figs. 4.14 and 4.16) causes the Neuron Chip to send a service pin message—identification number or Neuron ID—to the network. The Neuron ID is a worldwide unique 48-bit serial number assigned to the Neuron Chip by the manufacturer. This number is used to identify the node within and integrate the node into a network using LonWorks tools. The Neuron ID can also be entered manually or scanned in using a bar code reader. All LON devices have a label displaying the Neuron ID and a barcode.

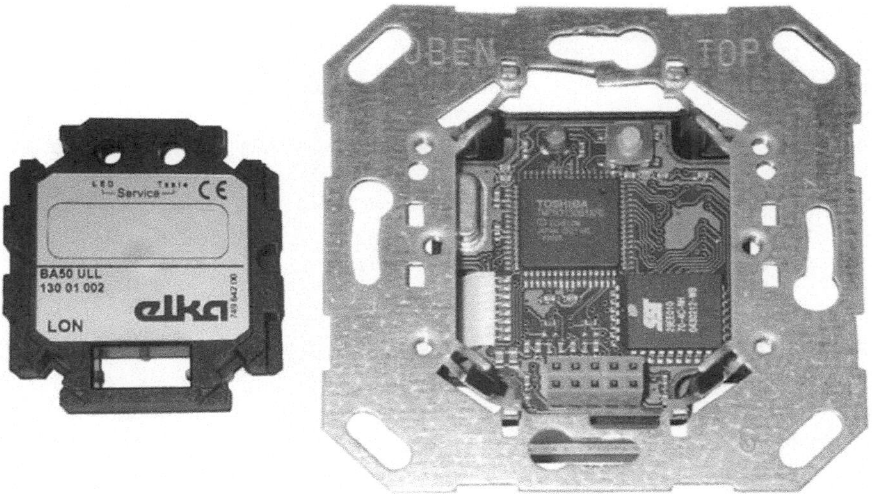

Fig. 4.16 A bus coupling unit with a Neuron Chip, memory, and a service LED and button [5]

4.4.2.5 Service LED

Almost all LON devices have a service LED (usually yellow) that shows the current state of the LON node during commissioning or troubleshooting. As soon as the devices power supply is switched on, the Neuron Chip in the device runs a self-test, the result of which is indicated by the service LED (see Table 4.1).

Some LON devices need to be customized for a particular function using an application program. A bus coupling unit, as shown in Fig. 4.16, can be fitted with a variety of different operating covers, which means it can be used as a light switch, for opening and closing blinds, or even as a setpoint adjuster. In each case, the operating cover's application program is loaded into the LON node. The service LED then shows whether the program has been loaded and whether the bus coupling unit has been connected to the network and can therefore communicate with other LON nodes.

During operation a so-called "watchdog" function constantly checks the status of the application. If the application is running as is should, a recurring signal stops the watchdog function. If an error occurs in the application, the signal stops and the watchdog is activated. The LON device shows that the watchdog function has started using the LED and, at the same time, resets and then restarts the application. If the error has not been fixed by resetting the application, the process is repeated, indicated by the LED error notifications shown in Table 4.1 [10].

Table 4.1 Service LED

Service LED's state	LON node's state
0.5 s on, then off	Node's application has been loaded and integrated into the network. Everything is ok
Flashes every 2 s	Node's application has been loaded, but has not been integrated into the network
1 s on, 2 s off, then on permanently	Application has not been loaded
Flashes every 0.8 s	Watchdog keeps resetting

4.4.2.6 Transceivers

A transceiver (Fig. 4.14) is used to connect a Neuron Chip to the network. A network interface connects the Chip to the transceiver and has a transfer rate of up to 1.25 Mbit/s.

Different types of transceivers ensure that the same device can be used on different transmission media. Some manufacturers build modular devices. This means that you just need to change the transceiver in order to use the device on a different medium, but the basic functionality of the Neuron Chip remains the same. Transceivers are either mounted on their own circuit board (Fig. 4.12) or encapsulated (Fig. 4.17). In both cases, you simply unplug the old transceiver and plug in the new one.

Table 4.2 shows the types of transceivers that allow LON devices to communicate with other devices over a physical network.

Fig. 4.17 An encapsulated FTT-10 A free topology transceiver [4]

Table 4.2 Transceivers for widely used media and network topologies

Medium	Transceiver	Transmission rate	Network topology	Network length	Power supply
Twisted pair	FTT-10A	78 kbit/s	Free topology, bus	500, 2700 m	Separate
Twisted pair	LPT-10	78 kbit/s	Free topology, bus	500, 2700 m	Via the bus
Twisted pair	TPT/XF-78	78 kbit/s	Bus	1400 m	Separate
Twisted pair	TPT/ XF-1250	1.25 Mbit/s	Bus	130 m	Separate
Power line	PLT-22	5 kbit/s	Free topology	[a]	Via a specialized power adaptor

[a]Depends on attenuation and interference

FTT-10A Free Topology Transceivers

LonWorks installations, particularly those in individual rooms, can be freely configured to fit a particular room or building. Free topology, as it is known, means that you can implement star, loop, linear and mesh topological structures as well as mixed configurations. Due to this high level of flexibility, however, a free topological network is limited to a range of 500 m. For a larger network of up to 2700 m, you must implement a bus (linear) topological structure. Twisted-pair cable is used, because it is relatively inexpensive.

An FTT-10A transceiver contains a transformer to galvanically isolate the LON device from the network. This is also means that components can be connected to the bus cable regardless of the polarity. However, if the network contains ring structures, the polarity of the network line must be adhered to. The LON node receives its power separately from the 12/24 V (DC/AC) voltage (common in the world of automation) or the power line. The transmission rate of 78 kbit/s is also fast enough to transfer analog signals.

LPT-10 Link Power Transceivers

The link power transceiver means that you do not need install a separate power supply. The LON device takes its power directly from the bus cable. A DC voltage of 42 V is superimposed on to the data AC voltage signal (Fig. 4.18). The device's

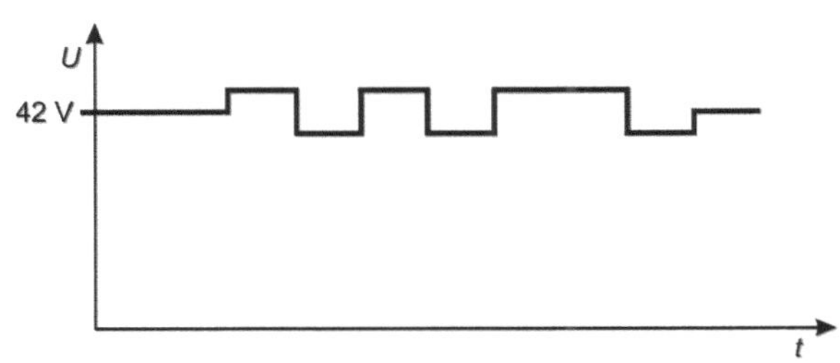

Fig. 4.18 The DC voltage superimposed on to the data signal

power adaptor is connected to the bus and supplies the device with the required operating voltage.

The transceiver separates the bus signal and the operating voltage, providing the LON node with an electric current of up to 100 mA from the network. This is enough to power, for example, heating valve actuators, LEDs and relays.

Free topology (FTT) and link power (LPT) transceivers can be used in the same network. The bus cable can be laid in a number of different ways. All topologies, except the ring architecture, are not sensitive to polarity.

Like FTTs, LTTs support a free topological architecture limited to a range of 500 m. For a larger network of up to 2700 m, a bus topology must be used. LPTs also use twisted-pair cable with a transfer rate of 78 kbit/s.

Smart-Transceiver FT 5000

The FT 5000 smart transceiver integrates a Neuron chip in a free topology transceiver (Fig. 4.19). It is compatible with both FTT and LPT transceivers. The combination of components allows the development of cheaper products and the user can save even more if he uses the LonMaker (described in Sect. 4.6.2.2) during installation in the LON network.

TPT/XF-78 and TPT/XF-1250 Twisted-Pair Transceivers

In addition to the transceivers mentioned in the previous sections, there are other types of transceivers that can be used with twisted-pair cable: the TPT/XF-78 and TPT/XF-1250 transceivers. These both have an in-built transformer that isolates them from the network, denoted by XF. The LON devices have separate power supplies.

These transceivers can only be used in a bus topology network that has not been expanded. The bus devices are connected to the bus via a short length of cable, or they are connected in a line and the signal passes from device to device. This allows them to communicate over long distances, but does mean that these transceivers are not suitable for connecting LON devices in a room.

The TPT/XF-1250 transceiver also has a very fast transmission rate of 1.25 Mbit/s, which means it is perfect for pathways that transfer a large amount of data, for example, pathways connecting high-performance DDCs to devices at the management level. The TPT/XF-1250 can also be used to connect the various LON

Fig. 4.19 FT 5000 smart-transceiver [4]

subnetworks over a router (see Sect. 4.5.1.3, Fig. 4.26) to form a loop or backbone network.

LON devices with twisted-pair transceivers (TPT) cannot be used with LON devices with free topology (FTT) or link power (LPT) transceivers, even though they use the same transmission medium.

PLT-22 Power Line Transceivers

Power line transceivers (PLT) provide an alternative to FTTs and LPTs particularly when refitting existing buildings. For example, a new energy meter installed in an old house would be fitted with a PLT (Fig. 4.20).

A PLT enables a device to transfer a signal over the existing power line. By sending the LON signal on the power line, you do not have to rip up floor boards and knock holes in walls to lay any additional bus cables.

The PLT-22 transceiver uses a frequency range from 125 to 140 kHz for transferring data. In the aforementioned transceivers, data is transferred over separate bus channels. With power line technology, however, other loads are connected to the power line as well as the LON devices. Signal noise can occur depending on the type of loads on the power line. For example, energy-saving lamps and traditional switch power adaptors can cause particular problems on the network. In addition, the electrical properties of the power line itself can cause signal attenuation. These factors mean that the maximum length of the network depends on the

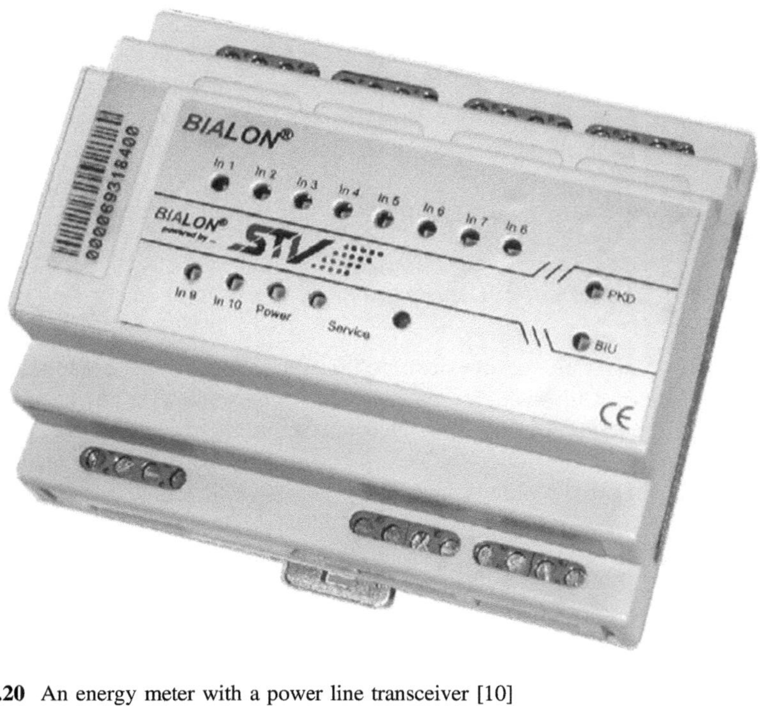

Fig. 4.20 An energy meter with a power line transceiver [10]

attenuation and the noise (interference). To reduce the amount of interference, the transmission rate is limited to 5 kbit/s.

The transceiver PLT-30 is another type of transceiver that works with the frequency reserved for European energy suppliers of 9–95 kHz. Especially electricity meters are often equipped with it to allow remote data requests.

Other Transceivers

Table 4.2 lists some of the most widely used transceivers. Other types of transceivers are also available for the following media:

- RS-485/EIA-485 twisted pair standard network
- 900 MHz and 2.4 GHz radio frequency
- 400–450 MHz radio frequency
- 1.25 Mbit/s coaxial cable
- Fiber-optic cable
- Infrared

Some new LON devices include a combination of neuron chip and transceiver, the respective products are labeled with FT 6000.

4.5 Transfer of Information Between LON Devices

This section will show you how to create a network from individual LON devices. The first step is to establish a physical connection between each device, taking into account the network topologies used for the transceivers mentioned in Sect. 4.4.2.6. The second step involves logically connecting the LON devices which means the devices can communicate with each other using the variables provided in the application programs.

4.5.1 Physical Network Topologies

A LON network has a unique physical architecture. The smallest component in a network is a Neuron Chip, which represents one LON node. If a device such as a powerful DDC contains a number of Neuron Chips, then each of these in-built Chips represents one LON node. The next section will explain the architecture of a LON network using free topology and link power transceivers.

4.5.1.1 Bus Networks

LON nodes that have a direct relationship with each other should be directly connected to each other. The greatest distance can be achieved using a bus topology as shown in Fig. 4.21.

A network with FTTs and LPTs can be up to 2700 m long depending on the type of cable used. The cable connecting the LON nodes to the bus must be no more than three meters in length.

To prevent the signal from being reflected back at the end of the cable, terminators (T) with a resistance value of R = 107 Ω are placed at both ends of the bus cable. The 42 V power adaptor required when using an LPT already has an in-built terminator.

It is hard to adhere to a bus topology in flush-mounted installations and is therefore hardly ever used in this situation. It is, however, often used in industrial applications to connect automation stations to the management level.

4.5.1.2 Star and Ring Networks

The star (Fig. 4.22a) and ring (Fig. 4.22b) topologies can be implemented when using free topology transceivers.

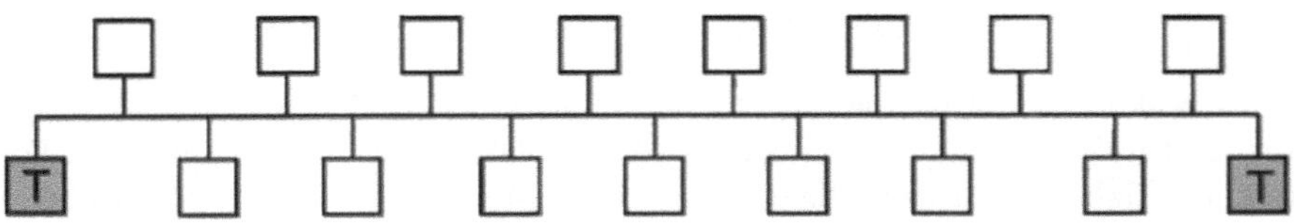

Fig. 4.21 LON nodes in bus topology with terminators

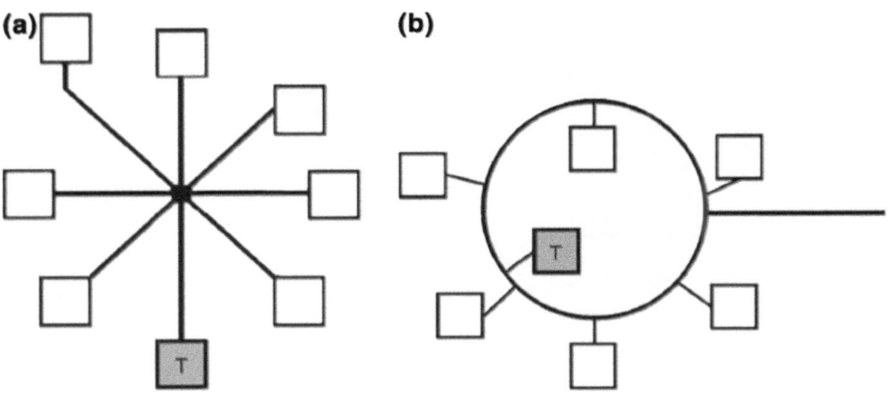

Fig. 4.22 Physically connected LON nodes in a star network (**a**) and a loop topology (**b**)

A star or ring (loop) network can have a maximum range of 500 m. The maximum distance between two LON nodes must be no more than 320 m, depending on the type of cable. A terminator with the value $R = 52.3\ \Omega$ is placed at the end of the network. If you are using LPTs, then a terminator is already built into the power adaptor.

A closed ring network is particularly suitable for the installation of LON devices in rooms. It allows you to refit components at a later date, without having to think about installation guidelines. What is important, however, is to observe the polarity of the bus channel. If you do not, then this could cause a short circuit, resulting in network failure.

4.5.1.3 Subnets

A subnetwork (subnet) represents the smallest part of a LON network. A subnet can contain a maximum of 128 addressable LON nodes (Fig. 4.23).

If the LON nodes are fitted with link power transceivers, these devices only place a small load on the bus because they have their own power adaptor. This means a subset can have a maximum of 128 nodes can.

Devices with free topology transceivers, however, put a greater load on the bus. This limits the number of LON devices in one segment to 64. In networks that have FTPs and LTPs, the maximum number of nodes allowed per segment is calculated by doubling the number of devices with FTPs and adding it to the number of devices with LPTs.

LON devices that operate directly with each other should be put in the same subnet. By doing this, you minimize the time it takes to execute a command. For example, light switches and the corresponding lamp actuators should always be in the same subnet.

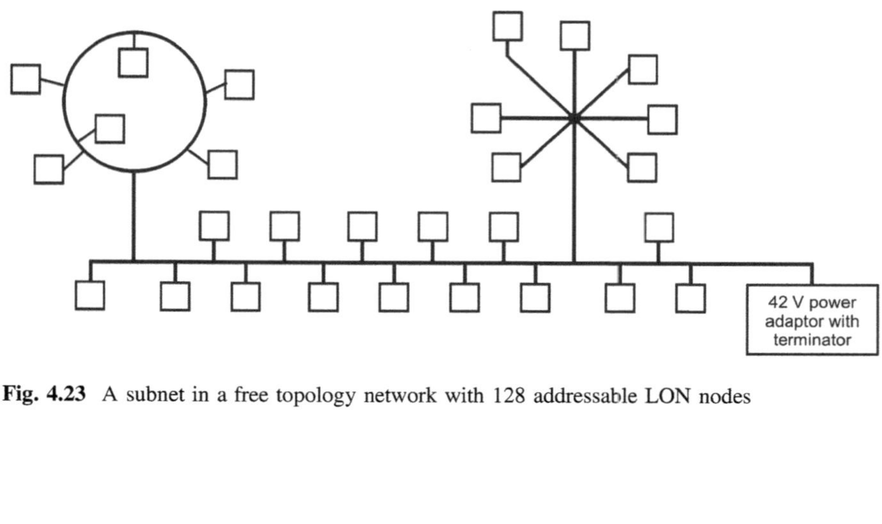

Fig. 4.23 A subnet in a free topology network with 128 addressable LON nodes

Repeaters

If you need to address more than 64 FTT devices within a subnet, then you will need a repeater (Fig. 4.24). You can also use a repeater when the segment exceeds the maximum length allowed.

A repeater connects two segments of the same medium and forwards, but does not filter valid data frames. A maximum of three repeaters can be installed in a row; more can cause signal delays, which can lead to communication problems.

Routers and Channels

If you want to use different transmission media in a LON network, then you must connect these media using a router (Fig. 4.25).

In LON terminology, the segments are known as channels. Routers can be used, for example to connect power line devices with FTT devices. You should note, however, that a router does not increase the number of nodes that can be addressed in a segment, it simply divides the various physical channels into different subnets.

Another difference between routers and repeaters is that a router can also filter data frames. It uses a routing table to determine whether a data frame is addressed to a node in the same network segment. It only forwards the data frame if the address is in the next segment. This way a number of subnets can be combined into one large network. The router counts as LON node.

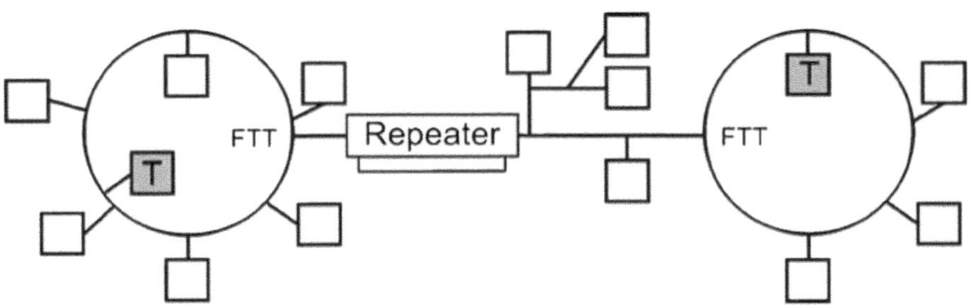

Fig. 4.24 A subnet with a repeater

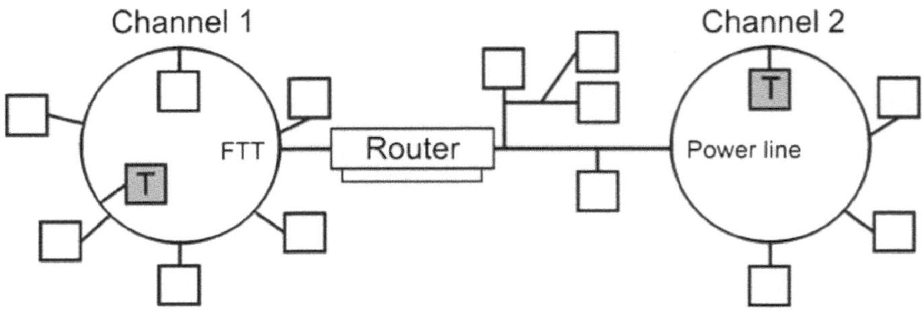

Fig. 4.25 A router connecting subnets and channels

4.5.1.4 Domains

If you have reached the limit of 128 LON nodes within a subnet, you will need to expand the network. This is done using a router, mentioned in Sect. 4.5.1.3. When expanding a network, you can choose to connect the subnets using a channel with a high transmission rate.

You should ideally use routers that have in-built TPT/XF-1250 transceivers, which enable a larger amount of data to be transferred over this backbone at a higher transmission rate (Fig. 4.26).

A domain consists of a number of subnets connected via routers. A maximum of 128 nodes can be addressed within a subnet using the LonTalk protocol. The router represents the 128th node in a subnet. A maximum of 255 subnets can be connected within a domain. If all the bus devices are in the same domain, then the domain information is not included in the address, resulting in shorter data frames and a faster user data transmission rate.

Within a domain, there are 255 subnets with up to 127 LON nodes, resulting in a maximum of 32,385 addressable devices.

If you should need more devices and therefore another domain, then you can connect the two domains using a high-performance router or using the control computer.

4.5.1.5 Creating the Network Structure

In order to create network structures, use the integration tools described in Sect. 4.6.2.2. In this case, the assignment of LON nodes to the subnets and domains described previously happens graphically.

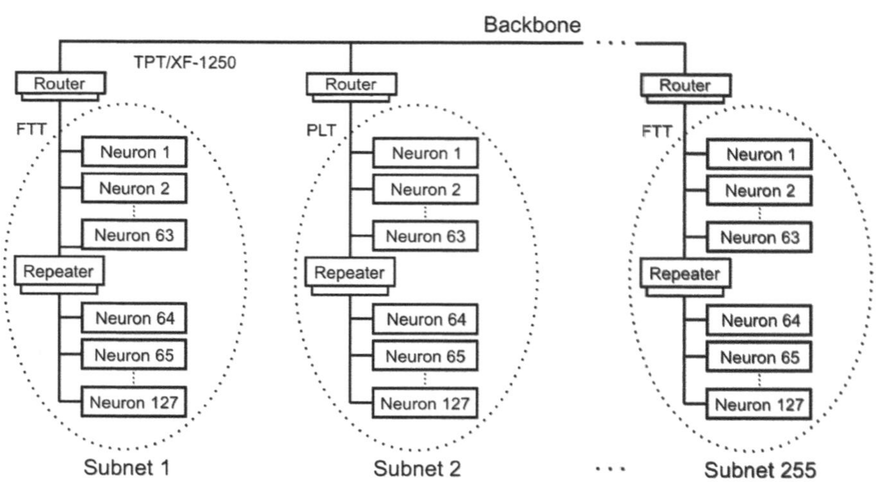

Fig. 4.26 Combining a number of subnets into a single domain

4.5.2 Frame Structure

For the user, the used frame structure is neither visible nor accessible during the creation of the network structure, as it is not required for a basic understanding of LON technology. Additional information on the topic, made for developers of LON devices or integration tools, can be found in [2, 3].

4.5.3 Media Access Control and Signal Coding

4.5.3.1 Predictive P-Persistent CSMA

In the LonTalk protocol, bus nodes (devices) that have equal access rights to the transmission channel use the predictive p-persistent Carrier Sense Multiple Access (CSMA) protocol to communicate with each other. All the nodes first listen to the transmission channel to see if a carrier (signal) is being transmitted by another node. Once the channel is idle, all the nodes must wait for a delay period to elapse. Once this delay has elapsed, each node then waits for a random delay period to elapse before transmitting (Fig. 4.27, top). The length of the delay and the data frame varies depending on the type of transceiver.

If no other LON node is accessing the network during a particular node's random delay, then this node starts transmitting. If two nodes have exactly the same random delay and want to start transmitting simultaneously, then both nodes sense this and stop transmitting. This can lead to considerable delays if there are a large number of nodes in the network.

One way of getting round this problem is to grant priority access to the channel (Fig. 4.27, bottom). Each node is assigned a set delay period according to its importance. The node no longer has to wait for the random delay, but can, if required, start transmitting during the priority window it has been assigned [10].

4.5.3.2 Differential Manchester Code

As mentioned in Sect. 4.4.2.6, FTT-10 and LPT-10 transceivers are polarity insensitive and can therefore be connected to the physical network regardless of the polarity. This is because the physical signals are encoded using the differential Manchester code.

The differential Manchester code creates a clock transition for each bit transmitted. The logical "0" is represented by an additional transition within the clocking period. This means that is does not matter whether a low or high signal is sent; if there is no additional transition, then it is interpreted as a logical "1" (Fig. 4.28).

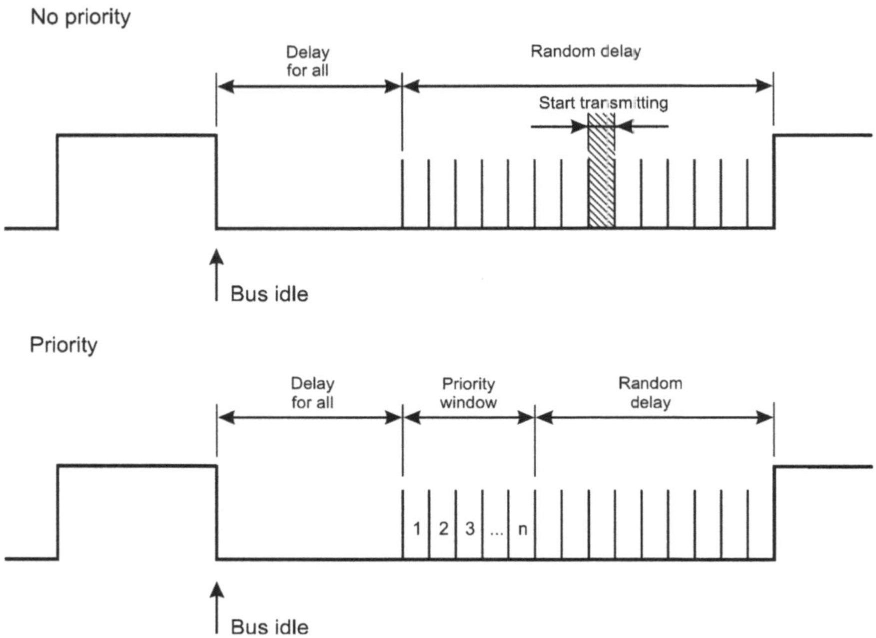

Fig. 4.27 Predictive p-persistent CSMA used by LON nodes to access the network

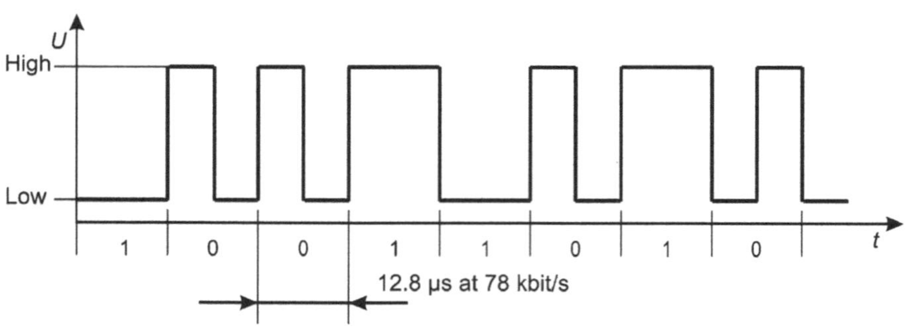

Fig. 4.28 The differential Manchester code

As a result, the number of transitions within a clocking period determines whether it is a "0" or a "1" bit. There is no dependency on the polarity of the bus channel. Another advantage of this method is that there is always a transition from high to low even when a series of logical "1 s" is being transmitted. This is particularly useful when synchronizing bus devices.

4.5.4 Logical Network Architecture with Network Variables

A physical connection between the LON nodes is a prerequisite to communication (see Sect. 4.5.1). The LON nodes exchange the actual data using network variables.

4.5.4.1 What Are Network Variables?

The functions a device is to have and the information it is to contain is determined during its development. A LON network in a building control system is a decentralized network with remote intelligent components. For the system to function properly, the individual LON nodes need to be able to communicate with each other over the network. Information is exchanged using network variables (nv). Figure 4.29 shows how a sensor and an actuator exchange information.

In this example, a sensor sends the current room temperature value to the heating valve actuator. To do this, an output network variable (nvo) is declared in the sensor. The output network variables for the sensor are defined when the sensor is developed. To find out the device's network variables, see the manufacturer's technical specifications. The selected variable contains the current room temperature value measured by the sensor. In the example above, nvo_roomtemperature is the output network variable.

The heating valve receives the actual room temperature value. The actuator's application program, supplied by the manufacturer, then compares this value with the desired room temperature value. The heating valve is adjusted (opened or closed further) depending on the difference between the actual and the desired temperature.

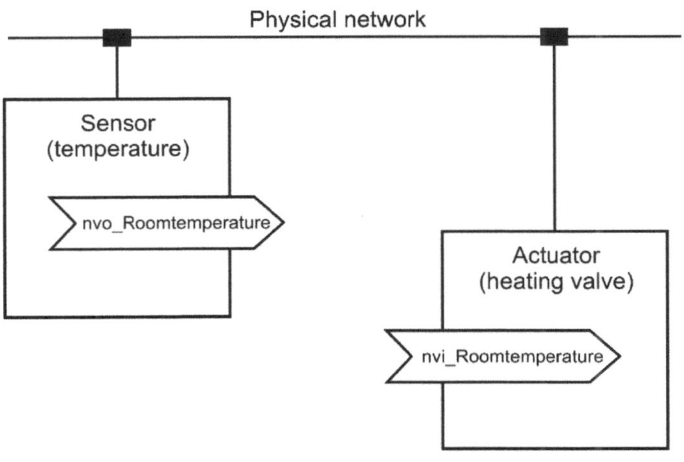

Fig. 4.29 Communication between LON nodes using network variables

In addition, the corresponding input network variable (nvo) needs to be declared in the actuator. To make sure you select the right variable, consult the device's technical specifications provided by the manufacturer. In the example above, nvi_roomtemperature is the input network variable.

Bear in mind that, particularly with sensors and actuators from different manufacturers, the variables may not have the same names. To select the corresponding variables, use the programming tools mentioned in Sect. 4.6.

4.5.4.2 Binding

After you have selected the right variables in both LON devices according to Sect. 4.5.4.1, you will need to connect them using a so-called "binding" tool. Binding these variables is carried out using the programming tools and is discussed later in Sect. 4.6 (Fig. 4.30).

Once the variables from the sensor and the actuator have been "bound", they are logically connected with each other. The binding tool automatically checks whether the variables match. If they do not, the binding process is cancelled. An output variable may only be bound to an input variable. The sensor then automatically forwards the new temperature value to the actuator. During commissioning, you can define how often a sensor transmits a value:

- Percentage change (only for analog values)
- Change of status (only for binary values)
- At regular intervals (heart beat).

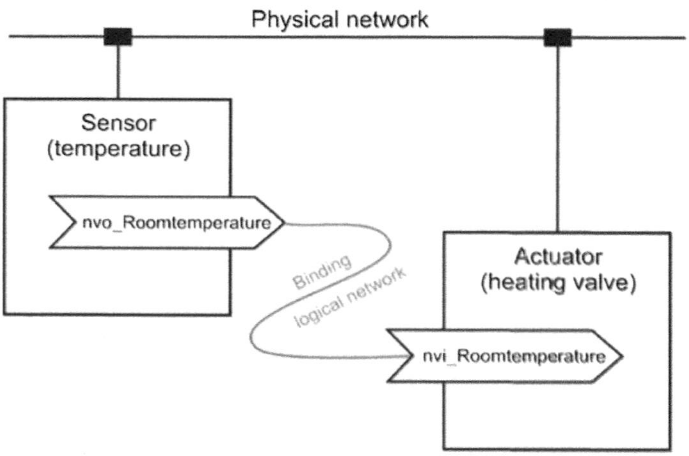

Fig. 4.30 Binding network variables

4.5.4.3 Acknowledgment Services

To prevent transmission errors, you can, when integrating LON nodes, define which transmission acknowledgment mechanism is to be used for network variable connection (Table 4.3).

When choosing the most suitable message service, you need to determine how important the message is, how many bus devices there are, and the number of data frames on the network.

- The *Unacknowledged* service is used to reduce the load on the bus. The recipient is not required to send confirmation; the data frame is only sent once, it is not re-sent. The drawback with this service, however, is that transmission errors can be detected.
- The *Unacknowledged repeated* service also does not require the recipient to send confirmation, but it does mean that the data frame is sent repeatedly. Temporary transmission errors, therefore, are not an issue but, as with the *Unacknowledged* service, longer-term errors are not recognized.
- The *Acknowledged* service is the standard setting for network connections. Every time a command is sent, all the addressees are required to send confirmation of receipt. Otherwise the data frame is re-sent repeatedly until all the recipients have confirmed that they have received it. This ensures successful data transmission. Bear in mind, however, that for a command broadcasted to several devices, each one will have to acknowledge receipt, which will cause a large amount of traffic on the bus.
- The *Request/Response* service is used for transmitting data to visualization and alarm systems. If an alarm system sends, for example, a request to one or several nodes, then the nodes only respond to this specific request. If no request is sent, then the nodes do not send any data.

4.5.5 Interoperability of LON Devices

The LonMark Interoperability Association has defined a set of guidelines for programming applications to ensure that devices from different manufacturers can

Table 4.3 Message services for logical network connections

Message service	Consequence
Unacknowledged	Data frame sent once but not acknowledged
Unacknowledged repeated	Data frame sent n-times but not acknowledged
Acknowledged	Data frame sent once and acknowledged by all addressed devices
Request/Response	Data frame sent once, answer contains the requested information

exchange data reliably. These guidelines are not mandatory, but not complying with them limits how the successful a device will be on the market because it won't be compatible with other products.

The guidelines are based on the following rules:

- Each LonMark node must have application-specific objects (functional blocks) and functional profiles with the predefined minimum scope of the application, In other words, a bus coupling unit that is being used as a light switch sensor must contain an object with a switch functional profile.
- The object must at least contain network variables specific to a particular application. For example, a light switch sensor must be able to send network variables for switching and dimming.
- Each LonMark node contains configuration parameters specific to a particular application. For example, with a light switch sensor you should be able to set the maximum output brightness value for a dimmer.
- All network variables used must be Standard Network Variable Types.

4.5.5.1 LonMark Objects and Functional Profiles

A LON device can be used for different purposes. The LON bus coupling unit [5] with the KNX operating panel [8], shown in Fig. 4.3, can be used as a lighting or blind controller or a setpoint adjuster.

If you want to use this LON device as a lighting controller, you first need to choose the appropriate operating panel and then select the object that corresponds to this operating panel in the software application preprogrammed into the device. The object's functional profile contains all the required input/output network variables and configuration properties. You will need to select the object when commissioning the LON node using the LonWorks tools described in Sect. 4.6. To ensure that devices from different manufacturers are interoperable, the LonMark Interoperability Association has assigned the objects specific functional profiles. Table 4.4 shows a few examples. For a complete list of all the functional profiles, visit www.lonmark.com.

Table 4.4 Functional profiles

Functional profile	Purpose
#0 Node object	Contains the LON device's basic functions and information
#3200 Switch	Used for all types of switches, control signal output from 0 to 100%, and also for dim functions
#3250 Scene panel	Used for triggering lighting scenes
#1060 Occupancy sensor	Occupancy(presence) button or sensor
#1040 Temperature sensor	Temperature value output
#8060 Thermostat	Controlling temperature with output for heating and cooling valves (actuators)
#3040 Lamp actuator	Coupling switch and dim actuators

If you wish to implement a lighting control system in a room, you will need to choose a light switch sensor and switch actuator from the manufacturer's product catalog. The product specifications list the functional profiles used. If the devices you choose are LonMark certified, they will contain the corresponding functional profiles.

For this lighting system the switch actuator has the #3040 *Lamp actuator* function profile (Fig. 4.31) and the light switch has the #3200 *Switch* functional profile (Fig. 4.32).

The network variables defined in the functional profiles are used to bind the light switch sensor and the lamp actuator together. This ensures a basic level of functionality even when using LON devices from different manufacturers.

4.5.5.2 Configuration Properties

The functional profiles for both the light switch sensor and the lamp actuator contain network variables and configuration properties. Figures 4.31 and 4.32 show optional network configuration inputs (nci).

When creating the application, the developer can define which configuration properties the device should have. The user can then customize the device's properties using an integration tool such as LonMaker.

Figure 4.33 shows the nci_MaxOut configuration properties being changed in LonMaker. The maximum brightness level for a dim actuator is set to 85%.

These configuration properties enable you to set other parameters such as for a dimming function, incremental dimming levels or a minimum brightness value.

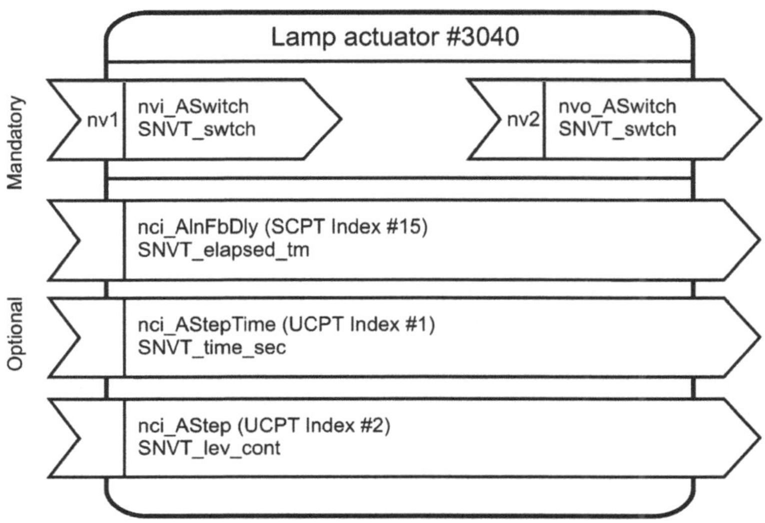

Fig. 4.31 The LonMark-standardized functional profile for a switch actuator [5]

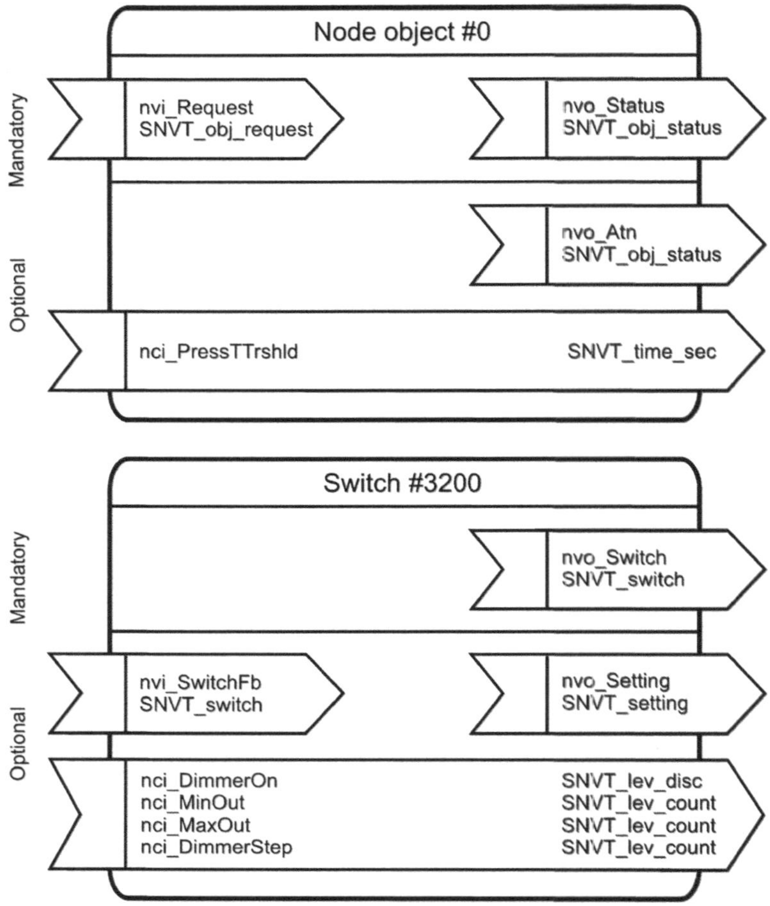

Fig. 4.32 LonMark-standardized functional profile for a light switch [5]

There are different types of configuration properties with which you can customize a device's properties:

- Network Configuration Variables (NVCs): These are used in the same way as network variables, which means they can also be reconfigured over the network. This is particularly useful when adjusting setpoint values.
- Configuration Properties (CPs): These are set inside the device itself using an integration tool and are saved in the database. This means they can still be accessed even if a node has been replaced.
- Standard Configuration Property Types (SCPTs): These are configuration properties that have been defined as standards by LonMark.

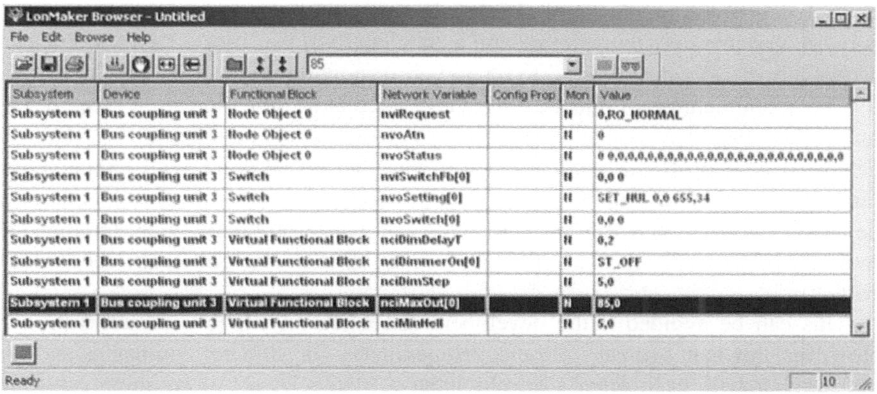

Fig. 4.33 Setting configuration parameters using LonMaker

Table 4.5 Example of a conflict arising over a network variable's range of values

Sensor manufacturer's temperature values (°C)	Values for the sensor manufacturer's network variables	Values for the actuator manufacturer's network variables	Actuator manufacturer's temperature values (°C)
0.0	0	0	0.00
5.0	50	50	0.50
10.0	100	100	1.00
21.5	215	215	2.15
50.0	500	500	5.00
215.5	2150	2150	21.55
550.0	5500	5500	55.00
6553.5	65,535	65,535	655.35

- User-defined Configuration Property Types (UCPTs): These properties, as the name suggests, are defined by the user and do not comply with the LonMark Standard.

4.5.5.3 Standard Network Variable Types in Building Automation

User-Defined Network Variables

The manufacturers of devices are completely free to choose and define their own network variables. If a developer of temperature sensors wishes to make the value available for the LON network, he or she must define the range of values and the degree of precision. In the example in Table 4.5 you can see that the developer decided on a temperature range of between 0 and 6553.5 °C. The sensor should have a degree of precision of 0.5 °C. For this the sensor developer defines a network variable nvo_temperature and a data length of two bytes. This variable can accept numerical values from 0 to 65,535 corresponding to $2^8 \times 2^8$ or 256×256.

The manufacturer of the heating valve (actuator), mentioned in Sect. 4.5.4, might, however, have other ideas as to what the parameters should be. He may decide to develop his actuator in such a way that it can work with temperatures ranging from 0 to 655.35 °C with a 0.01 °C degree of precision. For this he also defines a two-byte network variable. This will cause problems when binding the two network variables because they will not be able to interpret each other values correctly.

This can be avoided if the developers can agree on which parameters to use. However, this is not so easy if the developers work for different companies and are not familiar with each other's development activities. A set of common rules are therefore needed to ensure the interoperability of products from different vendors.

Standard Network Variable Types

The LonMark Interoperability Association has defined and published standardized system variables known as Standard Network Variable Types (SNVTs).

To avoid misunderstandings such as in the example above, the LonMark Association has defined a set of rules. The following definitions have been defined for the most commonly used variables:

- Area of application
- Name of the network variable type
- The variable's configuration
- The total length in bytes
- Value range, degree of precision and units

Table 4.6 shows a selection of the Standard Network Variable Types available.

The last two columns in Table 4.6 are particularly important. Defining the unit as well as the value range and degree of precision for a network variable, prevents the interoperability issues that occur with user-defined network variables [14].

Table 4.6 Example of the Standard Network Variable Types (SNVTs) used in building automation

Area of application	Type	Value range	Precision/unit
Dimming and switching	SNVT_switch	0 to 100 and 0/1	0.5% and off/on
Overall temperature	SNVT_temp	−274 to +6279.5	0.1 °C
Temperature for heating, cooling and ventilating	SNVT_temp_p	−273.17 to +327.66	0.01 °C
Overall percentage values	SNVT_lev_cont	0–100	0.5%
Brightness	SNVT_lux	0–65,535	1 lx
Flow rate	SNVT_flow	0–6553.5	0.1 l/s

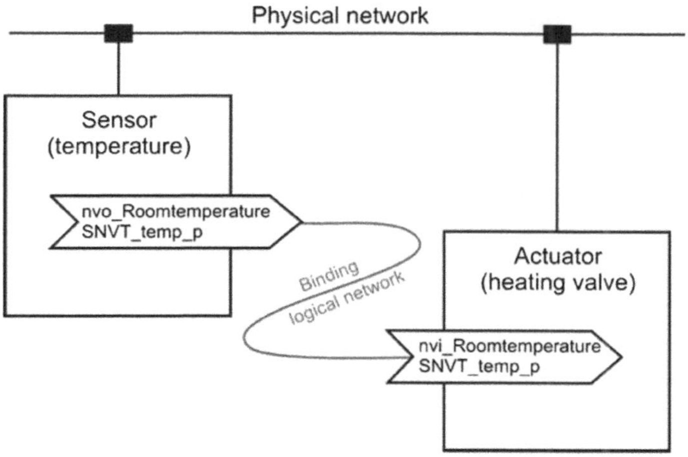

Fig. 4.34 Communication between LON nodes using Standard Network Variable Types

The temperature sensor and heating valve manufacturers make sure that each of their application programs has the appropriate functional profile (see Sect. 4.5.4.1). In the temperature sensor's functional profile #1040, the appropriate Standard Network Variable Type is used for the temperature output. According to Table 4.6 the SNVT_temp_p variable should be used in building control. If the heating valve developer has complied with the LonMark specifications, he should be using the #8060 profile and with it the same Standard Network Variable Type to receive the latest temperature reading over the network (Fig. 4.34).

So if both manufacturers have complied with the LonMark specifications, the heating valve application will then automatically interpret the value of 29,317 received via SNVT_temp_p as a room temperature of +20.00 °C.

The temperature reading requires a degree of precision of two decimal places, so that the actual value provided is more accurate than the device. At best a heating controller can be used to keep the temperature from deviating by more than 0.1 °C. A value that is less than that of absolute zero (−273.15 °C) is interpreted as a cable failure.

4.6 LonWorks Tools

LonWorks tools are used for programming Neuron Chips and integrating LON networks. In this section we will focus mainly on integration tools.

4.6.1 *Development Tools LonBuilder and NodeBuilder*

LonBuilder and NodeBuilder provide manufacturers of LON devices with a tool for programming Neuron Chips and a test environment. The development environment consists of hardware and software components that can be purchased from the company Echelon. Applications are programmed on a PC using the Neuron C programming language. NodeBuilder's integrated tools enable you to create LON-compliant devices directly. Furthermore, the software packet can be used to create the objects and functional profiles discussed in Sect. 4.5.5.1. Once the actual application has been permanently stored in the LON node's memory, it can no longer be changed by the user. The user is provided, however, with data in the form of a guideline template or with a plug-in that enables him or her to transfer the node's functionality to the network tool.

4.6.2 *Network Integration Tools*

Setting the LON node's parameters and integrating the variables from the device into the network is accomplished using special integration tools. There are a variety of tools from various manufacturers on the market. When equipping room automation systems (as mentioned in Sect. 1.6) with LON components, you need a large number of devices—the more devices the larger the network. If, at a later date, you want to expand the system using a different integration tool to the one you used initially, then it is important that you store the project data separately.

4.6.2.1 LonWorks Network Services

LonWorks Network Services (LNS) provide an integration-tool-independent database that operates according to the client-server principle. The server manages the central database system in which all network-relevant data is stored. All the devices' indices, the network variables, names and codes are stored in a standardized format. The integration tools work as the clients and are used to carry out graphical connections and bindings. The data is then transferred to the LON devices from the LNS database. It creates an independent interface between the integration tools and the hardware. Another advantage of this is that it allows several clients to access the server at the same time. This means, particularly with large projects, that several users can work simultaneously. You can also view the operating states in a building while changing an application's parameters. The server can either be accessed over the PC network or the Internet.

4.6.2.2 LonMaker Integration Tool

The most commonly used integration tool is LonMaker developed by Echelon. It enables you to design and set up a LON network with all the required network components, sensors and actuators. The integration computer is connected to the physical network using an external USB adaptor that acts as a gateway between the LON network and the computer's serial interface. The adaptor contains a Neuron Chip and therefore represents a node on the LON network (Fig. 4.35).

Another way of connecting the computer to the network is using an expansion card installed directly in the computer. A PCI expansion card is used for stationary systems and a PCMCIA (PC) card is used for transportable systems such as laptops. A third option is to provide the LON network with an Internet server through which the integration tools can access the network.

Unlike other programs, LonMaker has a graphical user interface environment, based on Microsoft Visio, that you can use to create the network and to bind the network variables (Fig. 4.36). To assign LON components to the subsystems described in Sect. 4.5.1.3, drag the SmartShape icon that represents the desired component to wherever you want it on the screen. Once you have positioned the SmartShape icon, the component's device template is loaded. A picture bar allows you to select the objects and functional profiles with all the pre-defined network variables provided by the manufacturer. To bind the variables, use the connector function icon on the picture bar. If these variables do not match, that is, they do not have the same Standard Network Variable Type, you will automatically receive an error message.

Figure 4.36 shows a light switch sensor and switch/dimming actuator assigned to one subnet. You can also see the gateway connecting the integration computer to the network. The line underneath the components represents the physical connection between the components. Above the components you can see functional

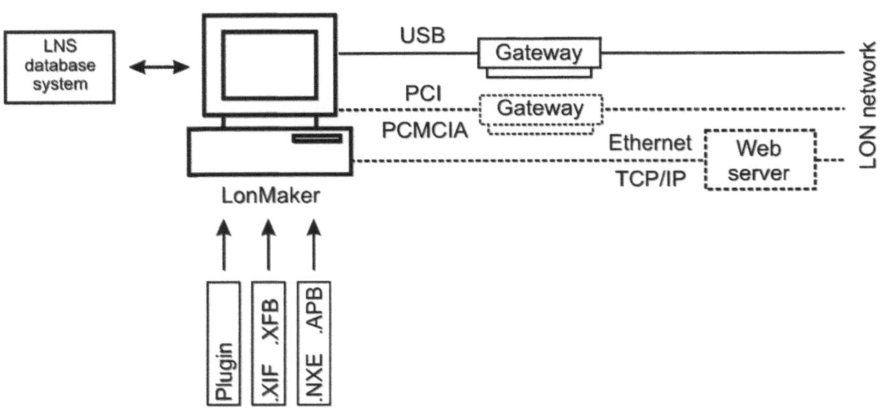

Fig. 4.35 Connecting an integration computer to a LON network

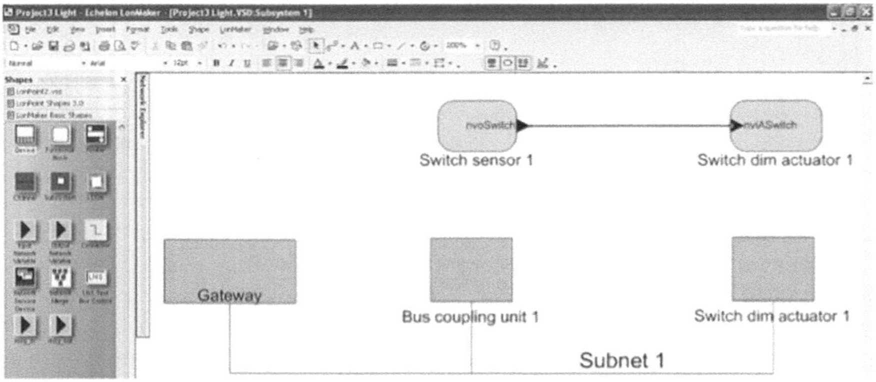

Fig. 4.36 Creating a project using LonMaker

profiles assigned to the objects. The binding (logical connection) between the variables (nvoSwitch und nviASwitch) for the selected Standard Network Variable Types in these profiles is represented by the dotted line. You can define the settings in the integration software so that the data is sent automatically to the components on the network.

The .XIF, .NXE and Plug-in Device Templates

To simplify the integration process, the LON device developer creates more files in addition to the application stored permanently in the LON device itself. The functional profiles and network variables in the application programs are created as device templates or plug-ins, and are either attached to the components or are available on the manufacturer's Web site. These additional files can be imported by the integration tools. This means that the properties and data of the relevant LON components are already available, which means they do not have to be entered additionally.

External interface file (.XIF) is the simplest format for this type of file. It contains all the data on the network variables in plain text. In addition, there is also the binary device interface file format (.XFB), which is a binary-coded compressed file imported by the integration program as a device template.

A more complex format is the .NXE file. It is loaded into all LON components that can run various application programs, for example, the bus coupling unit shown in Fig. 4.3, which can be used with a variety of operating panels and various applications. There is also a binary-coded, compressed version of this format that is read by the integration program as an .APB file. If needed there are programs that can create an identical version of an APB file from an .NXE file. Plug-ins are used in particularly complex applications such as heating controllers (Fig. 4.37).

If there are many control parameters, the manufacturers often provide additional programs to help you set these parameters. These programs are loaded as subprograms when the integration program is started. When you launch the integration

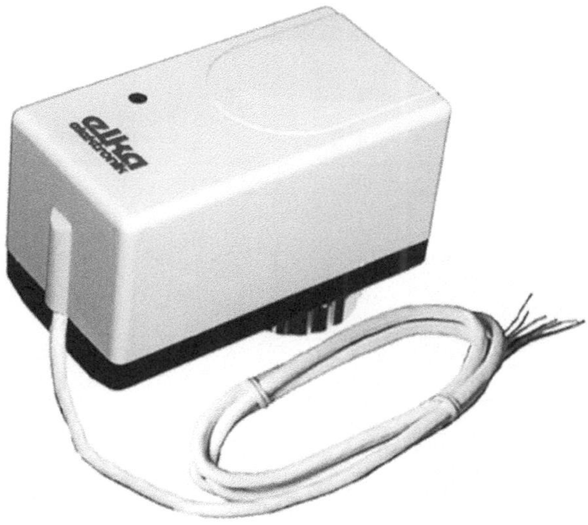

Fig. 4.37 A heating controller [5]

tool, these programs open in separate windows with their own user interface (Fig. 4.38).

Any changes made to the parameters using the plug-in are sent to the LON components by way of the LNS database system. You cannot use a plug-into select or connect network variables. You can only do this with an integration tool.

4.7 LonWorks System Architecture

In the previous sections, we introduced building automation and particularly building control solutions based on LON technology. LON-compatible DDCs (Sect. 4.2.2) are used for controlling and regulating building services. Whereas highly decentralized, intelligent components with their own functionality are predominantly used in room automation (Sect. 4.2.1).

4.7.1 Building Automation System with LON

In a building complex, the advantage of LON technology is that an open bus system can be used at all hierarchical levels, enabling you to create an open system architecture.

Figure 4.39 shows the typical architecture of a LON-based building automation system.

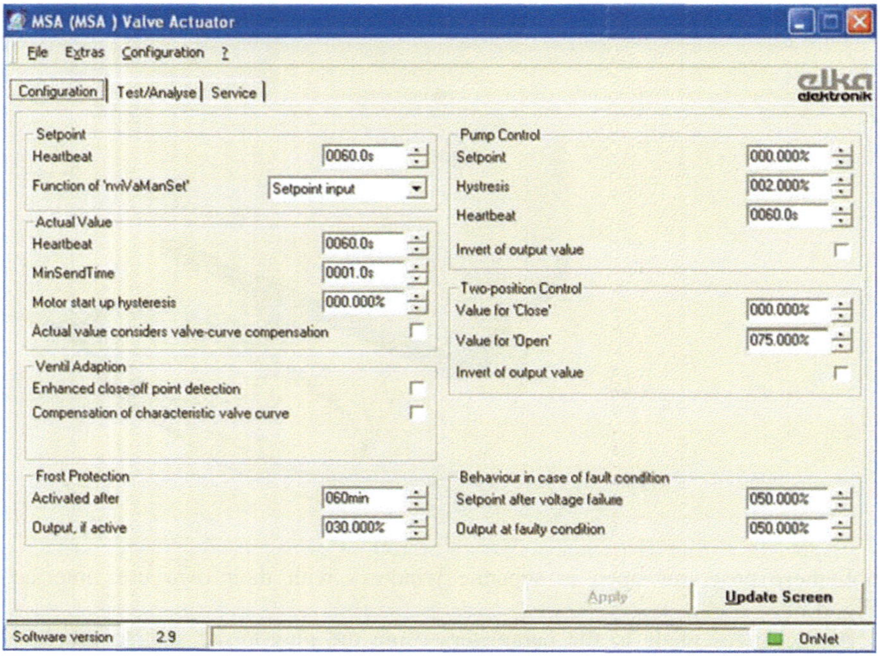

Fig. 4.38 An actuator for a heating valve with a control function and associated plug-in [5]

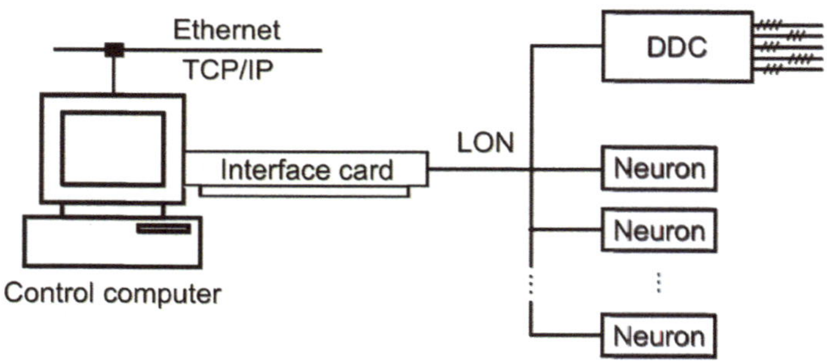

Fig. 4.39 Connecting a LON network to a control computer

The DDCs and system components used in room automation are grouped into subnets. The devices that are located nearest to each other, and which are logically connected, are usually placed in the same subnet. The control computer's in-built interface card enables the devices to communicate with the LON network. Depending on the control computer system used, you can also view the systems' operating states and measurement values. Furthermore, you can also set new set-points or run the energy management functions described in Sect. 1.4.3. Data exchange with other control computers occurs over the Ethernet connection shown.

4.7.2 Connecting LON Networks to the Internet

At the management level, you can implement Web-based solutions as well as the conventional control computers used in building automation systems. You can use the control computer shown in Fig. 4.39 as a Web server. There are many products currently on the market for this. In smaller LON networks, in particular, you can use a Web server instead of an control computer (Fig. 4.40).

A LON Web server (Fig. 4.41) has storage capacity (memory) for graphically displaying system states and event/alarm schedules and can run timer-switch programs, but it cannot execute energy management functions.

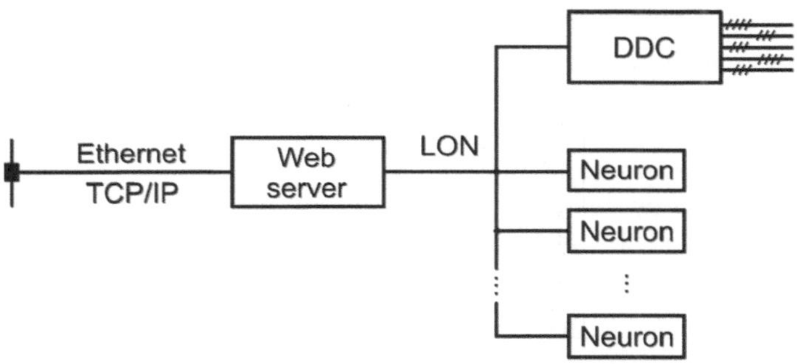

Fig. 4.40 Connecting a LON network directly to the Internet over a Web server

Fig. 4.41 The *i*.LON internet server [4]

The Web server's in-built Ethernet interface means that all system states can be accessed over the Internet. This is an alternative for private residential buildings and small buildings that do not have their own control centers.

4.8 Examples of Use

4.8.1 Lighting Control with LON

In our scenario, a conventional lighting control system in a room is to be replaced with LON components. To do this, you must first choose the devices for the project. A simple lighting control system requires the following devices:

- One bus coupling unit
- One operating panel as switches for the bus coupling unit
- One switch actuator with or without a dim function
- One power supply with terminator
- Three meters of twisted-pair cable
- One laptop installed with LonMaker
- One LON USB gateway

The operating panel is attached to the bus coupling unit. Depending on how the operating panel has been configured, the operating buttons can send either switch commands or dimmer values. Because you cannot just connect any operating unit to any bus coupling unit, you need to refer to the bus coupling unit's documentation to find out which operating unit to select. The developer of a bus coupling unit needs to consider the possible combinations when developing the unit, and must then provide the corresponding .XIF files or plug-ins. The command is executed by a switch actuator as introduced in Chap. 1 (Fig. 1.5). A twisted-pair cable connects the actuator to the bus coupling unit (see Fig. 4.42).

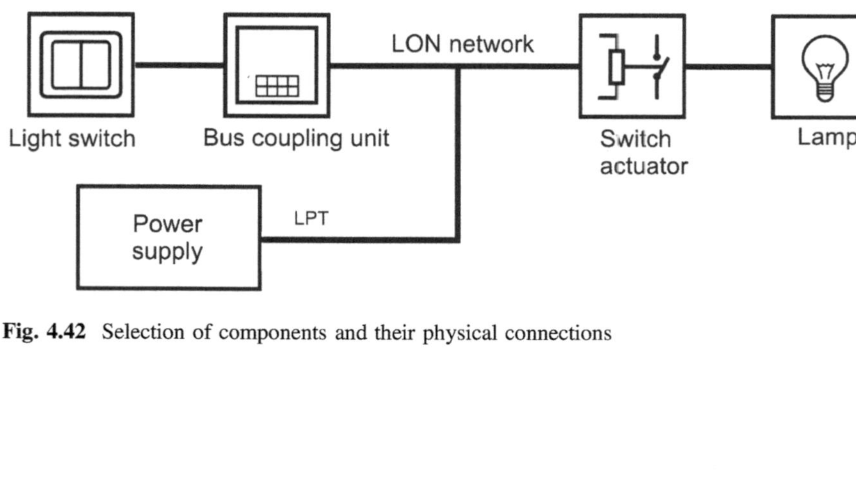

Fig. 4.42 Selection of components and their physical connections

You also need a power supply unit to supply power to the devices with LPTs that are connected to the LON network. The terminator required for the network segment is built into the power supply unit.

To prevent any problems when logically connecting the devices, make sure that all the devices are LonMark certified. If they are, the bus coupling unit will have an object with the functional profile #3200 *Switch* for the light switch, and the switch actuator will have an object with the functional profile #3040 *Lamp Actuator* (Fig. 4.43).

In the next step, use LonMaker to establish a connection with the LON network over a LON USB gateway. As discussed in Sect. 4.6.2.2, you need to assign the bus coupling unit and the switch actuator to a subnet and then select the required functional profiles using the graphical interface. Now declare an output variable for the bus coupling unit, in this case nvoSwitch, stored in the #3200 functional profile. Now select the corresponding input variable for the switch actuator, in this case nviASwitch. Finally, graphically bind the variables (Fig. 4.44).

You can select the dim function in the configuration parameters stored in the device template (see Fig. 4.33). If you have activated the dim function in this way, the switch actuator's output variable nvoASwitch should also be linked to the bus coupling unit's input variable nviSwitchFb, as shown in Fig. 4.44. This sends the current status of the switch actuator back to the light switch. This feedback is particularly useful if a lamp can be switched on or off from several dispersed switches. If the light has been dimmed, each switch is sent this updated status value (light has been dimmed). This means that when someone presses another switch to dim or brighten the lamp, a signal is sent based on this updated value and not the last value the individual switch actually sent. If switches did not receive an

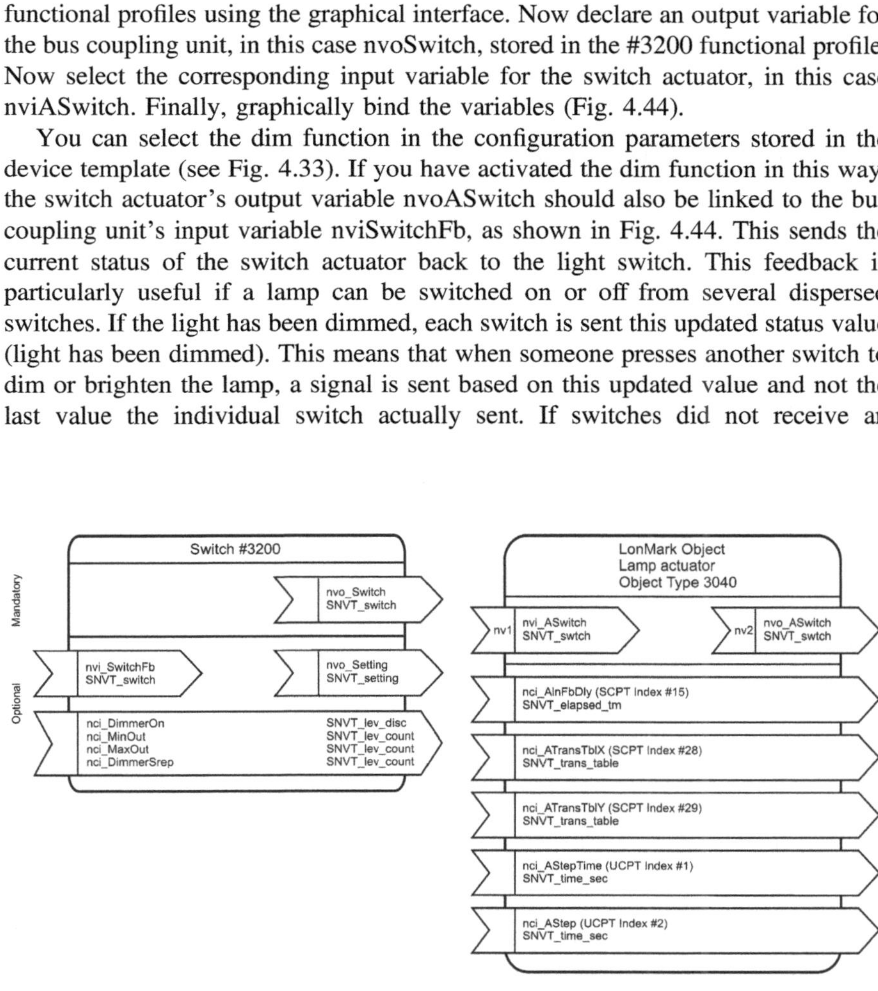

Fig. 4.43 The functional profiles #3200 *Switch* for the bus coupling unit and #3040 *Lamp actuator* for the switch actuator

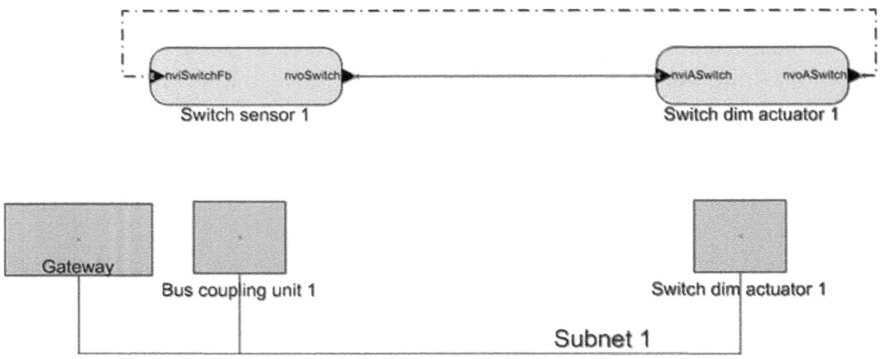

Fig. 4.44 An example of a lighting control system

automatic feedback, the lamp would jump from one state to another every time someone presses a switch, defeating the whole point of having a dimmer switch.

4.8.2 A Lighting Control System with a Panic Button Using LON

A particular advantage of bus technology is that it allows you to address several nodes simultaneously. This means that it is relatively simple to implement circuits that can switch several lights on or off from one light switch. This sort of arrangement can be used to increase security by enabling you to install a panic button. In the following example, the whole facility consists of two independent circuits for each light and an additional panic button. The following components are required:

- Three bus coupling units
- Three operating panels as switches for the bus coupling units
- Two switch actuators with or without a dimmer function
- One power supply with a terminator
- 5 m of twisted-pair cable
- One laptop installed with LonMaker
- One LON USB gateway

Figure 4.45 shows how the devices are connected to the physical network.

Light switch 1 should switch light 1 on or off, as in the previous example. In this example, there is also an additional light switch (2) and light (2). Bus coupling unit 3 is connected to an extra switch, the panic button, which switches both the lights on or off.

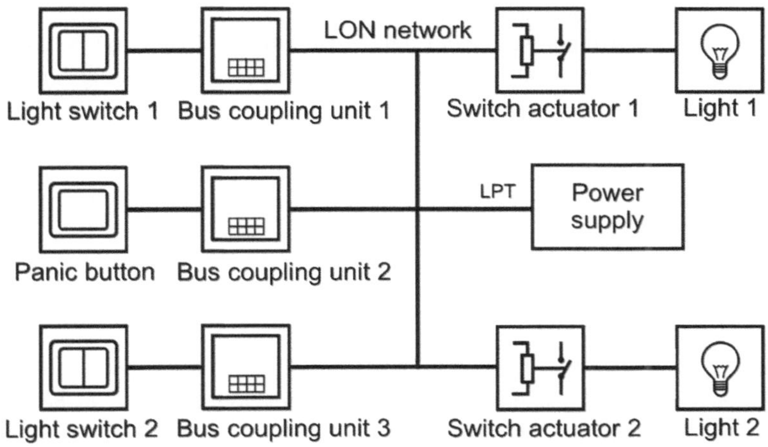

Fig. 4.45 Lighting control for two lights and one panic button

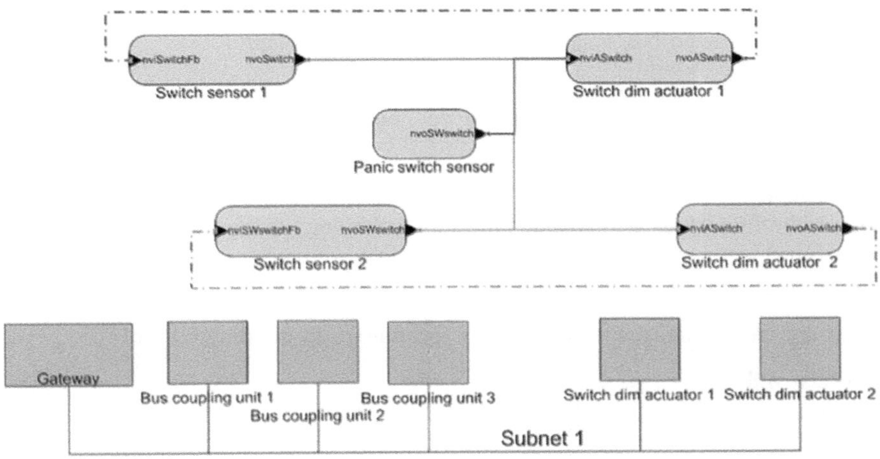

Fig. 4.46 Example of a lighting control circuit with a panic button

Figure 4.46 shows how the input and output variables from all the light switches and switch actuators are bound.

The logical connections (bindings) are the same as in the example in Sect. 4.8.1, but with an additional connection between the second light switch and the second actuator.

You now need to assign the bus coupling unit used as a panic button to the subnet and also select the functional profile #3200 *Switch*. Now declare the output variable nvoSwitch for the panic button. This output variable now needs to be connected to the lights 1 and 2:

Firstly, connect the output variable to input variable, nviASwitch, of light 1, and then connect it to the input variable of light 2. Both lights should now be connected to two output variables. Both variables can switch the corresponding lights on or off.

4.9 Exercises

Exercise 4.1
What does LON stand for?

Exercise 4.2
What is central control technology? What other developments had to be made to reach today's communication standards?

Exercise 4.3
What is a DDC?

Exercise 4.4
Which standardized bus systems are used in building control?

Exercise 4.5
Which building control functions can be implemented using LON technology? Give the generic terms.

Exercise 4.6
Sketch the functional principle of the LON technology using two lamps and three light switches.

Exercise 4.7
What are the advantages of using LON technology in building control?

Exercise 4.8
Which company developed LON technology?

Exercise 4.9
What is the role of the LonMark Interoperability Association?

Exercise 4.10
Which standards is LON technology defined in?

Exercise 4.11
What are the components of a LON node (device)?

Exercise 4.12
What is the service button on a LON device and what does it do?

Exercise 4.13
What are the main types of LON transceivers and what are their advantages and disadvantages?

Exercise 4.14
Design a physical LON network for an office building with 570 identical rooms, each with two windows and one door. Implement a presence-dependent lighting control system and heating and cooling functions. Include a 20% reserve for future additions.

Exercise 4.15
How can you connect two LON nodes without having to consider the polarity of the cable?

Exercise 4.16
What specifications must a LON device comply with to become LonMark certified?

Exercise 4.17
In LON technology what is meant by SNVT?

Exercise 4.18
Atwo-byte SNVT_temp_p network variable type has the binary value 0111 0010 1000 0100. What variable and unit does this correspond to?

Exercise 4.19
What are the LonWorks Network Services (LNS)?

Exercise 4.20
Design a LON network for a building complex with three central ventilation systems, a central heating and a central cooling system.
These systems together with the data from the power supply and sanitation systems are to be connected to a control computer.

The control computer must have remote access to the Internet.

Exercise 4.21
Draw all the components needed to control the temperature in a room with four static radiators. Then establish all the required logical connections based on the LonMark-certified functional profiles.

Literature

1. Busch-Jaeger Elektro (2009) Technische Unterlagen zu EIB/KNX-Geräten. Firma Busch-Jaeger Elektro, Lüdenscheid
2. Dietrich D, Fischer P (2001) LonWorks-Planerhandbuch. VDE, Berlin
3. Dietrich D, Loy D, Schweinzer H-J (eds) (1998) LON-Technologie. Hüthig, Heidelberg
4. www.echelon.com
5. ELKA (2006) Technisches Handbuch LON. Lüdenscheid

6. Gira, Radevormwald, Germany. www.gira.de
7. Hansemann T, Merz H (eds) (2003) Kommunikationssysteme für die Gebäudeautomation—Wirtschaftlicher Bedienungskomfort in Gebäuden mit Hilfe von Bussystemen. Aachen, Shaker
8. Jung, Schalksmühle, Germany. www.jung.de
9. LONMARK Deutschland e.V.: LON-Installationshandbuch, Berlin: VDE, 3. Auflage 2013
10. Meyer W, Stock G (2003) Praktische Gebäudeautomation mit LON. Hüthig & Pflaum, München, Heidelberg
11. STV Automation (2003) Zählererfassungsmodu. Firma STV Automation, Schloß-Holte
12. TAC (2002) Handbuch TAC Xenta® 280-300-401. Firma TAC, Malmö
13. Thermokon (2000) LonWorks-Technologie. Firma Thermokon, Mittenaar
14. Tiersch F (2001) Die LonWorks-Technologie. Desotron, Erfurt
15. Trox (2004) Automation und Systemtechnik LON-WA5/B. Firma Thermokon, Mittenaar

Chapter 5
BACnet

5.1 Introduction

BACnet (Building Automation and Control Network) is a standardized data communication protocol developed by the American Society of Heating, Re-frigeration and Air-Conditioning Engineers (ASHRAE) for use in building automation to enable devices and systems to exchange information. BACnet is used in numerous building automation systems worldwide and acquired the international ISO 16484-5 standard in 2003.

5.1.1 What Is BACnet?

BACnet evolved from the need for a standardized data communication protocol that would enable the various automation and control components in a building to communicate with each other, ensuring interoperability and manufacturer independence.

Before the introduction of BACnet, building automation was dominated by proprietary solutions from different manufacturers. With the result that a heating controller from vendor A would not communicate with the control center software from vendor B. Air conditioning, ventilation, lighting, and alarm systems, for example, were often planned separately and did not have interfaces over which they could communicate with each other or with a central control center.

Building developers were at a severe disadvantage. Either they had to do without being able to centrally control and monitor all these automation systems, and the synergies that this brought with it, or they were forced to purchase a complete solution from one vendor. When they wanted to expand an installation, they were

H. Merz et al., *Building Automation*, Signals and Communication Technology,
https://doi.org/10.1007/978-3-319-73223-7_5

effectively reliant on this one manufacturer and could not look around for potentially cheaper and better components elsewhere. BACnet, on the other hand, is an open multi-vendor communication protocol that allows components from different manufacturers to interoperate, providing increased market transparency and competition.

A similar development occurred in computer networks, notably with the advent of the Internet. Open protocols such as the Transmission Control Protocol/Internet Protocol (TCP/IP) have become accepted standards even though there are other proprietary protocols available. Today, the end user is able to choose hardware (network cards/adaptors, modems) and software (operating systems, application programs) components from a variety of manufacturers.

BACnet, as a vendor-neutral and license-free specification, enables you to interconnect all the systems in a building to form a functional network, increasing comfort, saving energy, improving security and reducing costs.

5.1.2 BACnet Organizations

Many users, manufacturers, institutions and advisors have formed the BACnet Interest Group Europe (BIG-EU). This interest group coordinates the BACnet activities in Europe and represents European interests in negotiations with ASHRAE [www.ashrae.org], the American manfucaturer association [www.bacnetassociation.org] and other international BACnet interest groups [e.g. www.big-na.org].

The BIG-EU publishes a vast amount of information on reference solutions and application examples in the BACnet Europe Journal. Additionally, it releases various literature on BACnet. The most notable publications are the current standard with its expansions and the guideline for interoperability in building automation based on DIN EN ISO 16484-5 [1]. This guideline together with "BACnet Gebäudeautomation" from Hans R. Kranz [2] are the most important references for detailed information on the topic.

5.1.3 Areas of Use

Commercial buildings are often extremely large. Each area in a commercial building also has different requirements with regard to heating, ventilation and air-conditioning. For this reason, the systems in the different areas and rooms in a building tend to be decentrally controlled by remote stations. The data (measurement values, operating states, alerts) from the remote (distributed) stations is

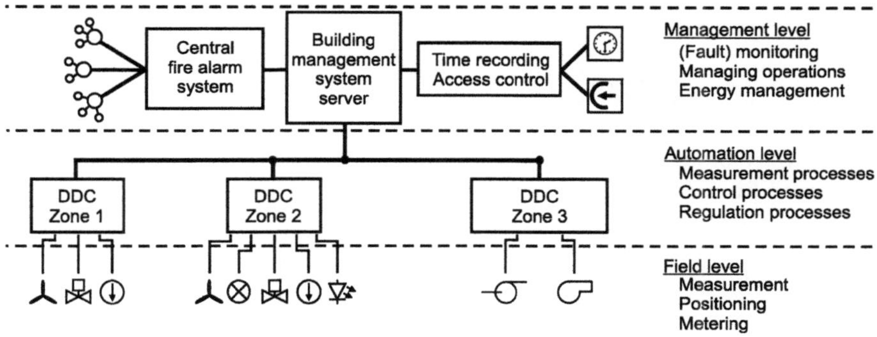

Fig. 5.1 Three-layer model in building automation

recorded by a control center, which has a graphical display, allowing for consistent cross-system access to all building data and control functions. In building automation this distributed system can be depicted using a three-layer model (Fig. 5.1).

The field level comprises individual sensors (temperature and switch sensors), actuators (control valves, drives, relays) and (room) control panels. These devices are connected to automation stations (DDCs) at the automation level, which in turn are connected to a building management system at the management level. From there you can monitor all sections of the network and also manage the individual systems and analyze faults. Clients in networks or office intranets, for example, can access the building management system server.

In general, process data at the field level does not need to be transmitted at a high rate. At the management level, on the other hand, the transmission rate must be considerably faster, because this level processes all data collected from all the systems. Response times, however, are not that important at the management level. It does not matter if it takes a few seconds for the operating status of a heating system to reach the control station. At the field level, however, response times must be quick. It should take no more than ten milliseconds for a signal to travel from a light switch to a lamp to turn it on or off. BACnet can be used at all levels of building automation, but is particularly suited for management functions. As a result, it is preferably used as a superordinate system in larger installations with LonWorks and KNX used at the field level.

Two of the most well-known BACnet projects are the United States Department of transportation building in Washington DC and the parliamentary buildings in Berlin, Germany [2].

5.1.4 Overview of the Basic Principles

5.1.4.1 Types of Information

The BACnet protocol defines how and which messages (data frames) can be transported from one device or system to another. The messages can contain the following information:

- Binary input and output values (e.g., "pump on/off," "window open/close")
- Analog input and output values (e.g., "current through temperature sensor"; "control voltage for valves")
- Software binary and analog input and output values (e.g., reading in °C from a temperature sensor)
- Schedule information
- Alarm and event information (e.g., movement detector, door contacts)
- Files (for securing configuration settings)
- Control logic.

5.1.4.2 Transport Ways

Messages are transported in various ways. A particularly good option is to use the existing local area network (LAN) used for the office intranet. BACnet messages can be transferred over this type of network, known as Ethernet, as well as over other local networks such as MS/TP, LonTalk or ARCNET. Point-to-point dial-up connections over a telephone line are suited for longer distances.

5.1.4.3 Objects

So-called "objects" are used to communicate with the individual devices and their functions within a BAC network. Imagine each building automation device as a set of data structures or objects. If a device has four digital outputs, two analog inputs and a controller, then the device must have an object for each of these components. Each object possesses certain properties (e.g., name and current status) that you can request or set. These objects enable you to receive information about a particular device without having to know anything about its internal set up or configuration—the objects are the key to interoperability. The same function can be implemented with different types of hardware and software. Standardized objects allow for multi-vendor access.

5.1.4.4 The BACnet Communication Architecture

BACnet is based on the Open System Interconnection (OSI) reference model (ISO 7498), which was developed for defining the architecture of communication systems [3]. Figure 5.2 shows the BACnet layers and the corresponding OSI layers.

The functions of the presentation, session and transport layers (layers 4–7 of the OSI model) are, as far as required, integrated into the BACnet application layer. A four-layer collapsed architecture was chosen to reduce the overheads, which generally increase the more layers there are, and to minimize the demands placed on the hardware and software for data transmission. The aim is to ensure the low-cost development and production of BACnet devices, which are installed with inexpensive microcontrollers that, in comparison with current PCs, have less—computing power and less memory. Furthermore, BACnet devices do not have hard drives or ventilation units. Without these components, BACnet devices have a greater lifespan but are less powerful.

What layers are required for a BAC network? BAC networks are predominantly local area networks that have external (Internet) interfaces. BAC devices are static, which means they always perform the same functions.

The physical layer provides a means of connecting BAC devices and transmitting the electronic signals that convey the data. The data link layer organizes the data into frames (or packets), regulates access to the medium, provides addressing, and handles some error recovery and flow control. These are all functions that are required in a BACnet system.

In a single network most network layer functions are either unnecessary or duplicate data link layer functions. For some BACnet systems, however, the network layer is a necessity, for instance, when two or more subnetworks are connected by a router using different data link layer options. In this case, there is a need to differentiate between local and global addresses and to route messages to the appropriate networks. The BACnet standard is designed so that devices are

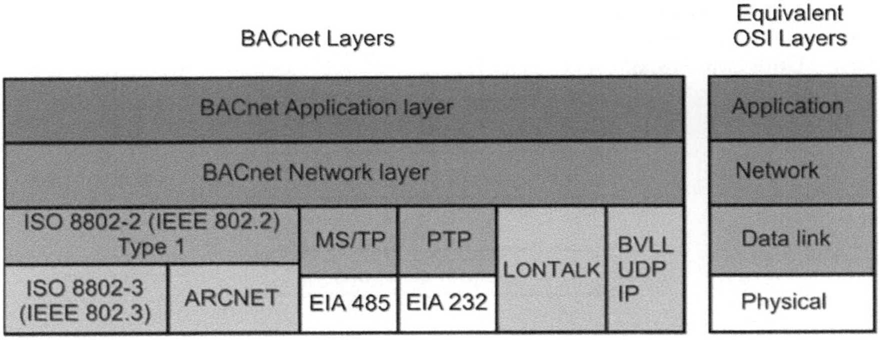

Fig. 5.2 BACnet layers compared with the OSI model

connected by only one logical pathway, which simplifies the whole routing process (compared to the Internet). The Internet has more of a partial mesh topology (see Chap. 2) with many alternative pathways. Compared with BACnet, the Internet requires far more complex routing algorithms to route packets via different pathways to recipients all around the world.

Most of the functions of the transport layer are similar to functions at the data link layer (e.g., data segmentation and flow control) but are different in scope. The data link layer focuses on point-to-point connections between two devices in a single network, whereas the transport layer deals with end-to-end connections across multiple networks. The functions of the transport layer are provided by the BACnet application layer.

Most communications in a BAC network are very brief and, as a result, do not require interrupt or restart mechanisms. In addition, formats do not need to be changed and data does not need to be compressed, which means that a BAC network does not require separate session and presentation layers.

For these reasons, a collapsed architecture made up of the physical, data link, network and presentation layers is the optimum solution for the simple and cheap implementation of BACnet in hardware and software components in building automation systems. In BACnet the layers shown in Fig. 5.3 encapsulate the data to create a message (data frame).

An application header (AH) is added to the application data (requests for manipulating objects) to form an application protocol data unit (APDU). This is then forwarded to the network layer which in turn adds a network header (NH) containing global network addresses. This network protocol data unit (NPDU) is then passed on to the data link layer, which adds a data link layer header (DLH) containing local network addresses. The physical layer then transfers it over the transmission medium. This encapsulation process is reversed at the recipient's side.

The following sections will explain the associated protocols and transmission methods.

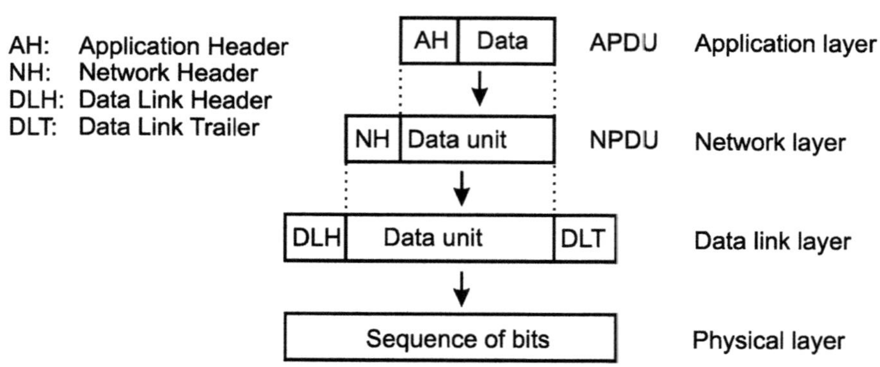

Fig. 5.3 Encapsulation of BACnet data

5.2 Physical Layer and Data Link Layer

With BACnet, data can be transferred over different types of networks. BACnet supports the local area network (LAN) technologies shown in Fig. 5.4.

These technologies vary depending on performance, cost and type of transmission media such as twisted-pair, coaxial and fiber-optic cable. The following criteria are often used when choosing a suitable LAN technology for a BAC network:

- Transfer rate: note that the actual data throughput is often considerably less than the transfer rate. This is because the addition of extra information (addresses, error detection, etc.) reduces the net data throughput.
- Response time: the time it takes for a command to be sent and the action to be carried. With non-deterministic transmission technologies such as Ethernet, the duration cannot be predicted but is often so small that it is insignificant.
- Number of devices (nodes)
- Maximum cable length
- Cost.

The following sections will explain the main features of the various transmission technologies.

5.2.1 Master-Slave/Token-Passing

The Master-Slave/Token-Passing (MS/TP) protocol is a simple and inexpensive technology, particularly suited for smaller control and operating units that do not need a fast transfer rate.

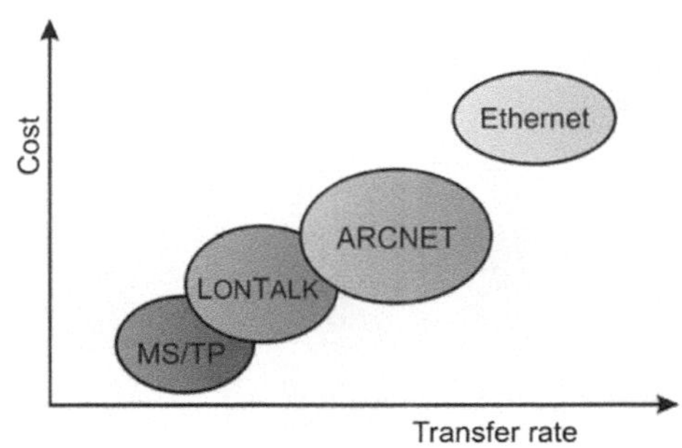

Fig. 5.4 LAN technologies for BACnet

5.2.1.1 Physical Layer (EIA-485 and EIA-232)

An MS/TP network uses shielded twisted-pair cable with a recommended maximum segment length of 1200 m. The physical layer is based on the EIA-485 standard (RS-485). An EIA-485 connection represents serial data transmission, in other words, the bits are transmitted one after another (Fig. 5.5).

The data is transferred in the form of characters. One character represents eight bits. A start bit signals the start of a character and one or two stop bits signal the end of a character (one stop bit for MS/TP). The interval between two characters can be any length, which is why this is also called asynchronous data transfer. The standard transfer rate for MS/TP is 9600 Bd. Device manufacturers, however, also support 19,200, 38,400, or 76,800 Bd. The low transfer rate and the simple protocol mean that MS/TP can be implemented in low-cost microcontrollers.

There are two significant differences between EIA-485 and the well-known PC serial interface EIA-232 (RS-232): The EIA-232 is ground-referenced voltage interface, in other words, the bits are converted into voltage levels, whereby a voltage between −3 and −15 V compared to ground represents a logical "1"; and a voltage between +3 and +15 V represents a logical "0." In the simplest case, you therefore only need three cables to connect two devices (Fig. 5.6).

EIA-485, on the other hand, is a differential voltage interface. This means that only the recipient can interpret the voltage difference (Fig. 5.7).

The transmitter applies a voltage $U_{AB} = \pm 3$ to 6 V between cables A and B. A positive voltage U_{AB} represents a logical "1", and a negative voltage represents a logical "0".

Interference can occur on long cables and in the vicinity of electrical loads. This can lead to an increase in potential ΔU on both cables. The voltage difference U_{AB},

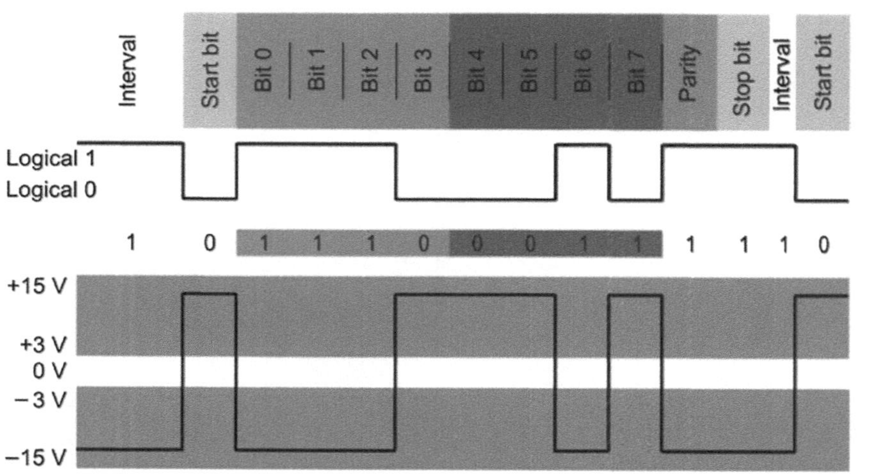

Fig. 5.5 Asynchronous character transfer using EIA-485 and EIA-232 as an example. The logical states assigned to the voltage on the channel only apply for EIA-232

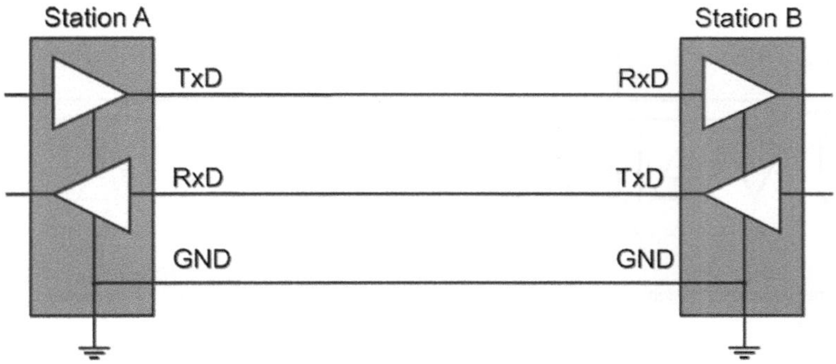

Fig. 5.6 EIA-232 (RS-232) with three cables: TxD (transmit data), RxD (receive data) and GND (ground)

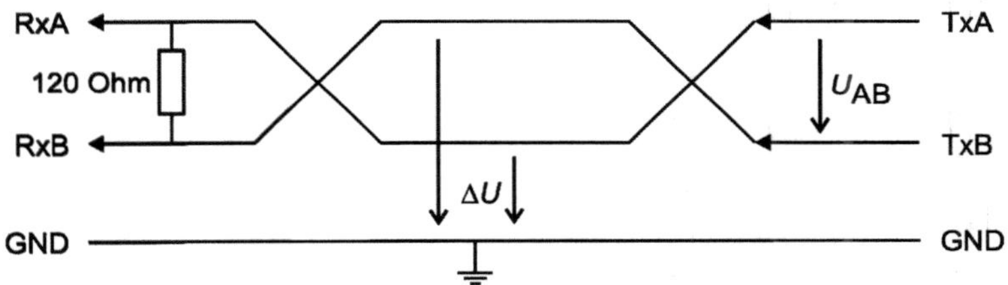

Fig. 5.7 EIA-485 (RS-485) with three cables (half duplex)

however, remains unaffected by this so-called common-mode interference. This makes the EIA-485 interface particularly resistant to interference. It allows you to use long cables and is, therefore, widely used in industrial communication technology. EIA-232 is only used for short distances, less than 15 m, with a typical transfer rate of 9600 Bd to 115 kBd.

Another difference between EIA-485 and EIA-232 is that the latter supports full-duplex point-to-point connections, in other words, the simultaneous transfer of data in both directions. Whereas EIA-485 only allows half-duplex transmission, that is, transmission in alternating directions. As a multi-point bus system, EIA-485 can have upto 32 nodes per segment (Fig. 5.8).

Each segment must be properly terminated. This involves attaching resistors that correspond to the wave impedance on the cable to stop the data signals from being reflected at the open ends of the cables.

An electrical signal sent by a transmitter travels along the cable at a typical propagation rate of approximately two-thirds of the speed of light. According the law of conservation of energy, which states that energy cannot be created or destroyed but can change its form, if a signal reaches the end of a cable, it will be completely reflected (this also occurs with sound waves). The reflected signal can superimpose itself on the transmitted signal and cause interference at the receiver end. To prevent this from happening, the ends of the cable are terminated with a

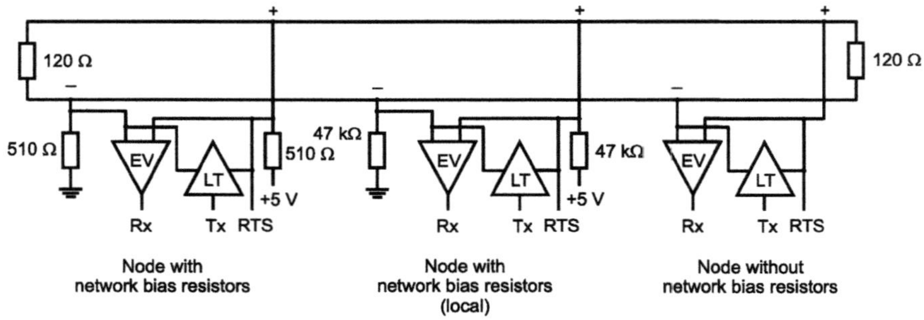

Fig. 5.8 An EIA-485 network showing three types of nodes

resistor. This converts the incoming signal energy into heat energy, stopping it from being reflected. The resistor chosen must match the wave impedance on the cable. Wave impedance is one of the attributes of a cable (not dependent on the length of the cable) measured in ohms, but is not to be confused with the resistance of a copper cable. The wave impedance is in fact the ratio of an electromagnetic wave's voltage and current used to transport data on a cable. The product of the voltage and current is the signal energy. If the termination impedance equals the wave impedance, then the (signal) energy is converted into heat without any of it reflecting.

Each MS/TP EIA-485 segment is also fitted with bias resistors (Fig. 5.8). Without these resistors the differential voltage equals approximately zero when the channel is idle, that is, nothing is being transmitted. Noise on the cable can cause the differential voltage to fluctuate gently between positive and negative values. The stations connected to the cable may then falsely assume that data is being transmitted. During the interval phase, it is therefore better to simulate a fixed voltage difference on the cable using a resistance network (see voltage divider in Fig. 5.8: +5 V—510 Ω—terminators each 120–510 Ω—GND). Each station can also be fitted with an additional local bias resistor. The received signals are then strengthened by an input repeater and the sent data by a line driver.

5.2.1.2 Data Link Layer

At MS/TP physical layer the individual characters are constructed into frames, according to the following format (Fig. 5.9).

The preamble signals the start of the frame. For the token-passing protocol, the frame type shows whether the frame contains data or control information.

2 octets	1 octet	1 octet	1 octet	2 octets	1 octet		2 octets
Preamble	Frame type	Destination address	Source address	Length	Header CRC	Data	Data CRC

Fig. 5.9 MS/TP frame format

The source and destination addresses each take up one octet. The destination address 255 is reserved for broadcasts. The other 254 addresses can be assigned to nodes. The length portion is reserved for data and is two octets long; however, only values between 0 and 501 are allowed. The frame header and the data portion are each secured by their own CRC checksum.

The data link layer also controls access to the transmission medium. With token-passing an "authorization to send" (token) is forwarded from one station to the next. As soon as a station receives the token, it is then allowed to communicate with other stations. A MS/TP network has master and slave stations. Only a master station can receive a token and initiate data exchange. Slave stations do not receive a token and simply wait for requests. BACnet field devices (e.g., sensors and actuators) represent the slaves that are connected to a master such as an automation station over MS/TP. If there are two or more masters, then the master with the token must forward it after a specific amount of time has elapsed. This means that you can calculate exactly when a master station will receive the token. This method is therefore known as a deterministic media access control method.

5.2.2 *Point-to-Point*

Two BACnet devices (half-routers) can exchange messages over an EIA-232 point-to-point connection (PTP) (Fig. 5.10). Remote areas that do not have physical Internet access via, for example, DSL can be reached over a dial-up connection using a modem.

5.2.2.1 Physical Layer

The BACnet standard does not define the means by which the physical connection is established. It only specifies data link layer protocols that enable the reliable transfer of data frames over an established physical layer connection. It is worth noting, however, that although a PTP connection is capable of full duplex operation, it is usually only temporarily available and has a slower transfer rate.

The standard does not define security mechanisms such as password requests. Many modems do, however, have an automatic callback option. When this option is activated, the modem immediately disconnects the call once a connection has been established and then calls back on a preset number. This prevents an

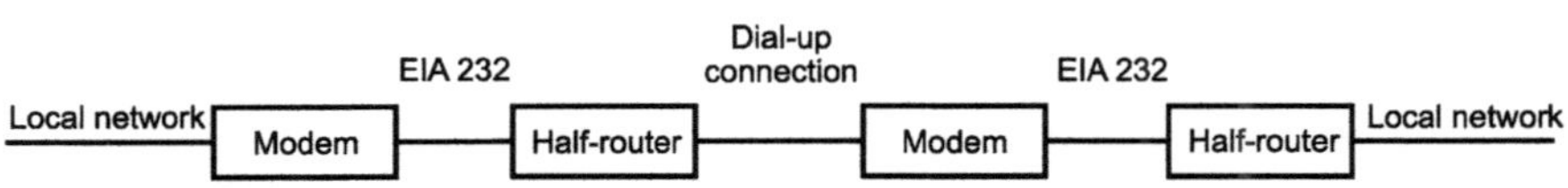

Fig. 5.10 A point-to-point dial-up connection with two half-routers

unauthenticated caller from using the modem from a different telephone extension. Some modems have an option that combines both automatic callback and password request.

5.2.2.2 Data Link Layer

Once the stations have been physically connected, they exchange control frames to establish a data-link-layer connection. The stations can then transmit BACnet data frames until the data-link-layer or the physical connection is terminated. As only two stations are involved in a point-to-point connection, the frames do not have an address. The preamble signals the beginning of the frame (Fig. 5.11).

The frame type shows whether the message contains data or control information. For example, a *Connect Request* message establishes and a *Disconnect Request* message terminates the connection. The answering stations then reply by sending either a *Connect Response* or a *Disconnect Response* message. One feature of the PTP protocol is that each data frame is confirmed by a special control frame. As with MS/TP, the frame contains the length of the data and the CRC checksums protect the integrity of the frame header and the data.

5.2.3 *Ethernet*

Ethernet is the most widespread LAN technology. It was first used in office intranet systems but has now found a niche in industrial communication systems and wide area networks. Nowadays office and industrial buildings are generally interconnected using Ethernet, so it was only a matter of before it would be used in building automation. Ethernet comprises layers one and two of the OSI model. Higher protocols include TCP/IP in intranets and BACnet in building automation. Ethernet was originally developed in the early 1970s and is standardized as IEEE 802.3. Ethernet has evolved over the years, most notably shown by the huge increase in its bit rate (10 Mbit/s, 100 Mbit/s, 1 Gbit/s, 10 Gbit/s) and its extensive downward compatibility, the availability of transfer media such as fiber-optic cable, and the introduction of wireless LAN technology.

2 Octets	1 Octet	2 Octets	1 Octet		2 Octets
Preamble	Frame type	Length	Header CRC	Data	Data CRC

Fig. 5.11 Point-to-point frame format

5.2.3.1 Transfer Using Twisted Pair

Twisted Pair

Twisted-pair cable comprises a number of twisted pairs of copper wires. The simplest form comprises two pairs (four wires), one pair for transmitting (Tx) the signal and the other for receiving it (Rx) (Fig. 5.12).

Twisting the wires reduces electromagnetic interference from external sources, caused by crosstalk and electromagnetic induction from neighboring pairs of wires. Each pair of wires and the whole cable can also be shielded with a sheath of metal. You should choose a cable that meets the data transmission requirements. As a rule, the higher the transfer rate of a digital signal, the higher the frequency. This affects attenuation and crosstalk between the pairs of wires. Attenuation refers to the reduction in the strength of a signal as it travels from the transmitter to the receiver. This is caused by the resistance (Ω) of the metal wires and dielectric loss in the insulation material. These factors both increase the higher the frequency; so that the higher the transfer rate the shorter the cable is allowed to be.

Crosstalk occurs when a signal transmitted on one pair of wires causes an undesired effect on a neighboring pair of wires. This is caused by capacitive and inductive coupling between neighboring pairs of wires because of overlapping electrical and magnetic fields. This effect also increases the higher the frequency and means that each pair of wires must be shielded. As a result, attenuation and crosstalk limit the maximum possible rate data can be transferred.

Transmission Standards: 100Base-TX, 1000Base-T

Table 5.1 shows the Ethernet standards most commonly used. The 100 Base-TX is normally used to connect workstation computers to a network, whereas 1000Base-T is used for connecting network components such as switches and routers or for connecting servers.

100Base-TX is often referred to as "fast Ethernet," because it is ten times faster than its predecessor 10Base-T. This faster transfer rate is thanks to a special transmission procedure known as multi-level transmission (MLT). In MLT three

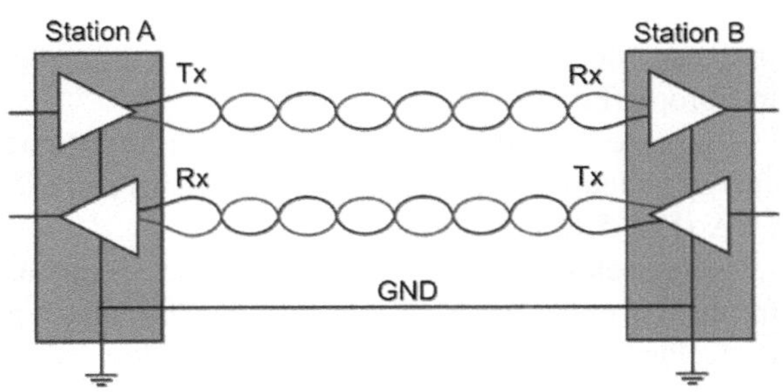

Fig. 5.12 Twisted-pair cable for transmitting and receiving data

Table 5.1 Types of twisted pair used with Ethernet

Name	Speed	Maximum cable length (m)
100Base-TX	100 Mbit/s	100
1000Base-T	1 Gbit/s	100

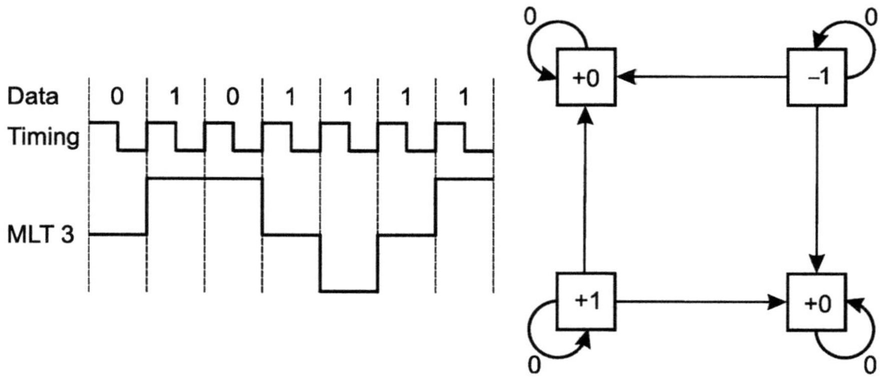

Fig. 5.13 MLT-3 transmission (example of a signal and a status diagram)

values are sent, in other words, there are three possible symbols (−1, 0, +1) or output states that can be transmitted (Fig. 5.13).

With a logical 0, the output signal stays at the previous level, whereas with a logical 1, the signal moves up or down. The advantage of this method is that the voltage level changes slowly (it does not jump from +1 to −1) and therefore the signal's frequency is comparatively low. The lower the frequency the lower the attenuation and crosstalk between neighboring wires. For this reason, with fast Ethernet you can use cables up to 100 m in length. The cables usually have eight wires, of which only four are actually used (two for Tx and two for Rx). The other wires can be used for transmitting telephone signals, known as cable sharing. To allow for an upgrade to Gigabit Ethernet, you should avoid using all eight wires.

When selecting a cable, you can refer to the EIA/TIA-568 (Electronic Industries Association/Telecommunications Industry Association) classification, which defines cable classes and their associated attributes. For fast Ethernet you should use a category 5 cable that fulfils the minimum requirements for crosstalk and frequency-dependent attenuation. Screened twisted pair (ScTP) is commonly used in Europe (Fig. 5.14).

The outer sheath reduces interference from external signals and unwanted signal energy radiation. Whereas unshielded twisted pair (UTP) is more widely used in the USA (Fig. 5.15).

Nevertheless, UTP is cheaper than ScTP and has a smaller diameter, which means it is easier to install. The best type of twisted-pair cable for transmitting data is shielded twisted pair (STP) or screened shielded twisted pair (SsTP) (Fig. 5.16).

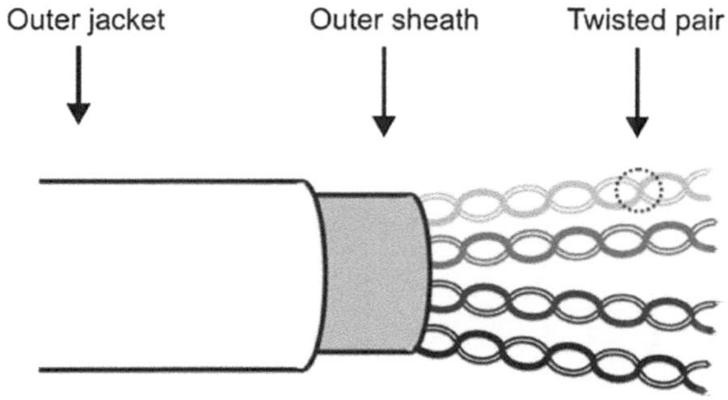

Fig. 5.14 Screened twisted pair

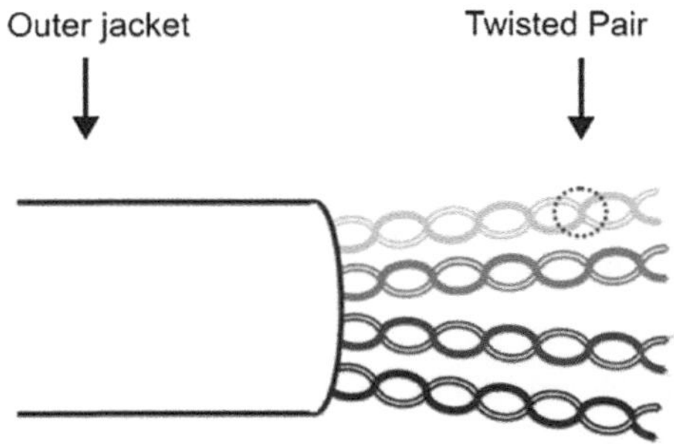

Fig. 5.15 Unshielded twisted pair

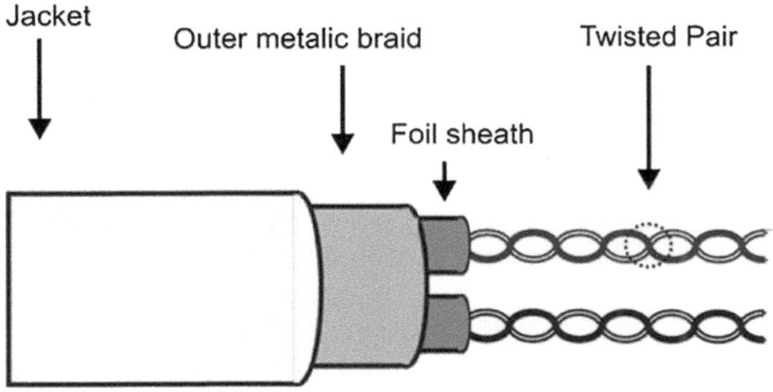

Fig. 5.16 Screened shielded twisted pair

This type of twisted-pair cable reduces interference from external signals and also helps prevent crosstalk between neighboring pairs of wires. SsTP, however, is expensive and difficult to install and, as a result, is seldom used.

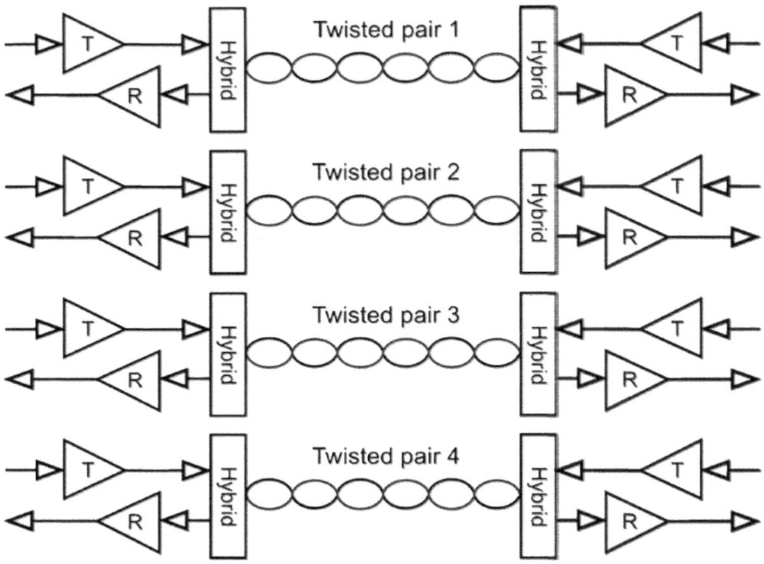

Fig. 5.17 Using all four pairs of wires for 1000Base-T (1 Gbit/s) over twisted pair

The continuing developments in Ethernet technology through to the faster transmission rates have led to a demand for higher-performance cable, as shown by the new categories 6 and 7. With 1000Base-T you can just about get away with using category 5 or 5e for transmitting 1 Gbit/s over a distance of up to 100 m. You need to narrow the frequency spectrum and thus limit the attenuation and crosstalk. You can do this by using all four pairs of wires simultaneously, reducing the transmission rate to 250 Mbit/s for each pair. The data flows on the wires pairs in both directions simultaneously, which means you need a hybrid to separate the data when it arrives at its destination (Fig. 5.17).

With 1000Base-T you can also use five symbols (signal levels) and therefore reduce the transmission rate still further. Using superior cable enables you to increase the transmission rate above and beyond 1 Gbit/s. A 10 Gbit/s twisted-pair transmission standard has now been developed.

Autonegotiation

Most network cards and network components support more than one transfer rate. In addition, older devices may be used that only support the 10 Mbit/s standard 10Base-T. Autonegotiation is an Ethernet procedure that enables two devices to choose the best transmission speed, duplex mode (full or half) or standard supported by both devices, which saves the user having to set the connection parameters manually.

After a cable connection has been established between two stations, each station sends a normal link pulse (NLP) at regular intervals. With NLPs, a station can only detect the presence of a connection to another station. In an improved version of autonegotiation, a station sends a burst of pulses, known as a fast link pulse (FLP), which allows stations to exchange information and choose the best common transmission parameters (speed, full or half duplex). If necessary, you can overwrite these selected parameters manually in each device.

Autosensing

Most of the latest network components also have autosensing capabilities, enabling them to recognize the type of cable automatically. Straight-through cable is used to connect network cards and network components such as switches (Fig. 5.18).

The pins are connected over the wires 1:1. This ensures that the transmitted signals (TD+, TD−) arrive at the corresponding receiver inputs (RD+, RD−) on the other side. To connect two stations that are of the same type (network card to network card, switch to switch), you need to use a crossover cable so that the transmitters are connected with their corresponding receivers. With autosensing, however, the type of cable used is no longer an issue because the parameters are set automatically.

Power-Over Ethernet

Twisted-pair cable can also be used to supply the connected devices with power. This procedure, known as "power-over Ethernet", saves you having to install a separate power supply and therefore reduces the amount of cable and space needed. It can also be turned on or off remotely. Network components, such as switches,

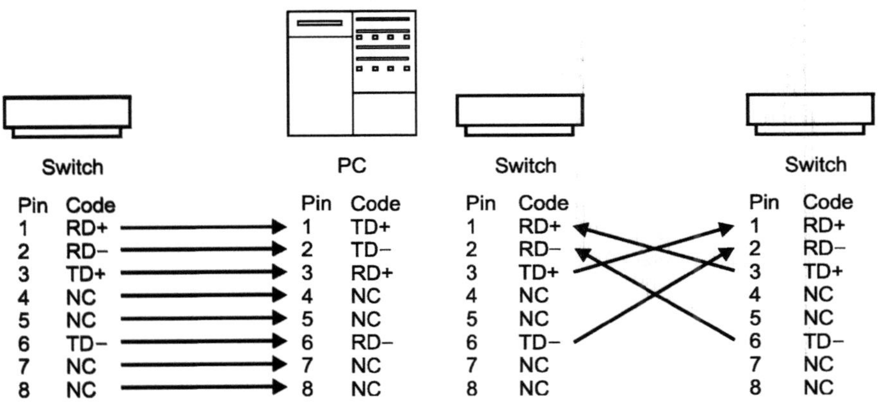

Fig. 5.18 Straight-through cabling (left) and cross-over cabling (right)

usually already have power-over Ethernet capabilities and can supply the connected devices with up to 10 W. The power can be supplied over any unused wires (e.g., in fast Ethernet) or the DC voltage can be superimposed on the data signals.

5.2.3.2 Network Components (Repeaters, Bridges, Hubs and Switches)

The original version of the Ethernet uses coaxial cable to connect all stations to a local area network (Fig. 5.19).

A station can only send data if the transmission medium is idle (that is no data is being transmitted). When there is a low load on the network, collisions can occasionally occur if two stations start to transmit at the same time. In other words, both signals overlap and interfere with each other. Network cards listen to the channel continuously, enabling them to detect a collision and stop the transmission in time. After a randomly selected interval, the frames that collided are re-sent. This is known as carrier sense multiple access/collision detection (CSMA/CD). CSMA/CD works very well when only a few stations want to transmit data. If a large number of stations all want to send data at the same, then this can cause a lot of collisions and can even lead to network failure. For this reason, modern cable-bound local area networks no longer use CSMA/CD, but work with point-to-point full duplex connections over switches.

For a better understanding of CSMA, it is worth learning about the historical development of CSMA and the associated topologies and network components. CSMA methods are, however, experiencing a resurgence, notably in wireless LANs. After all air is a shared medium on which collisions may occur.

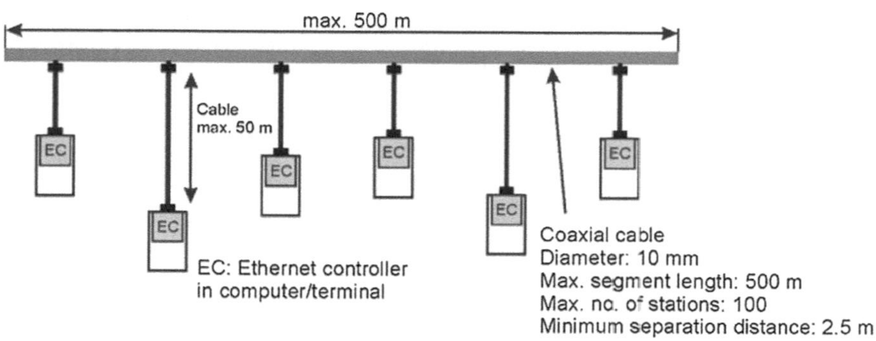

Fig. 5.19 Original version of the Ethernet (10Base5) with coaxial cable

Repeaters

The maximum length of a network segment depends particularly on the attenuation of the electrical signal on the cable. You can install a repeater between two segments to regenerate and forward the signal and enlarge the network (Fig. 5.20).

A repeater is a layer one (physical layer) component. It forwards electrical signals, but cannot interpret bits or the content of frames. One of its drawbacks, however, is that it regenerates all incoming signals. It even forwards a signal sent from one station to another in the same network segment to all network segments that are connected to the repeater. This causes an unnecessary load on the segments and therefore limits the maximum number of stations that can be included in the whole network. Because a repeater forwards all incoming signals, collisions can also occur in different segments (remote collision) as well as between two stations in the same segment (local collision). As a result, all network segments that are connected by a repeater are a collision domain.

Bridges

Bridges are data link layer devices that are used to divide a network into smaller collision domains. A bridge links two network segments and decides whether to forward data frames from one segment to the other (Fig. 5.21).

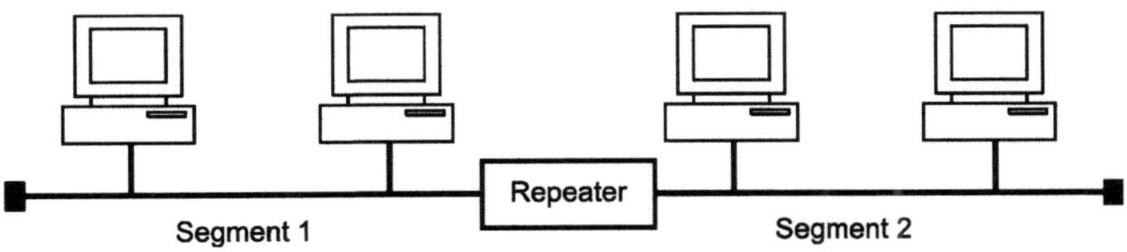

Fig. 5.20 Two segments and a repeater

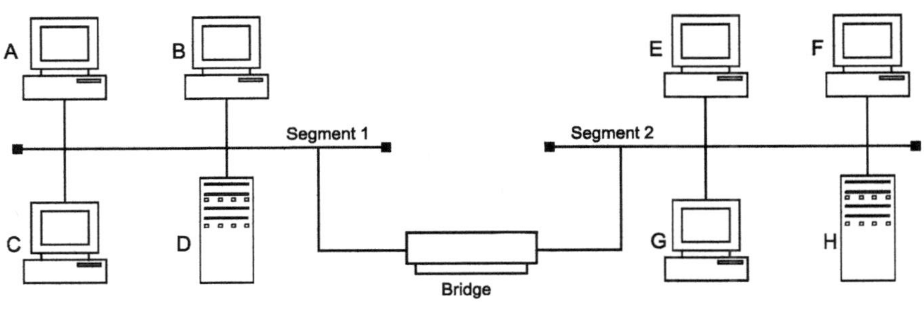

Fig. 5.21 A bridge connecting two segments

The bridge only forwards a data frame when the destination station is in the other segment. For this a bridge uses an address table containing the MAC addresses of the devices and which segment they are in. Once the bridge is switched on, this address table is automatically completed without the need for an administrator (Table 5.2). When a data frame arrives at a bridge, the sender's MAC address and port are entered in the address table. The bridge then checks whether the recipient's MAC address is already in the address table. If it is, then the bridge only forwards the data frame out of the port the recipient's segment is connected to. If the recipient is not on this segment, the frame is discarded. If the MAC address is not in the table, then the bridge forwards the data frame to the next segment (out of the other port) as a precaution.

Broadcast frames are an exception. Broadcasts are always forwarded by the bridge, because they are addressed to all the stations in a network. As a result, network segments that are connected by a bridge are called a broadcast domain.

Hubs

The original Ethernet networks with coaxial cable were prone to errors (disruption of the bus cable paralyzed the network segment) and were expensive due to use of the special transceivers for routing the signals from the bus to the individual stations. Star networks were developed for this reason, and coaxial cable was replaced with the cheaper alternative of twisted-pair cable. At the center of the network was a new network component, the hub. A hub is a multiport repeater. It is a layer-one device that regenerates all incoming signals and forwards them to all other ports. A hub and the stations connected to it make up a large collision domain (Fig. 5.22).

Instead of a computer (station), you can also connect another hub to the system, increasing the number of connections available. All connected stations must,

Table 5.2 A bridge's address table (lists for segments 1 and 2), as shown in Fig. 5.21

Computer's actions	Bridge's actions	Segment 1 list	Segment 2 list
A sends a frame to B	A is entered in the segment 1 list. The frame is also sent to segment 2	A	–
B answers A	B is entered in the segment 1 list. The frame is not forwarded to segment 2	A, B	–
B sends a frame to G	The frame is also sent to segment 2	A, B	–
G answers B	G is entered in the segment 2 list. The frame is also sent to segment 1	A, B	G
F sends a frame to E	F is entered in the segment 2 list. The frame is also sent to segment 1	A, B	G, F
E answers F	E is entered in the segment 2 list. The frame is not sent to segment 1	A, B	G, F, E

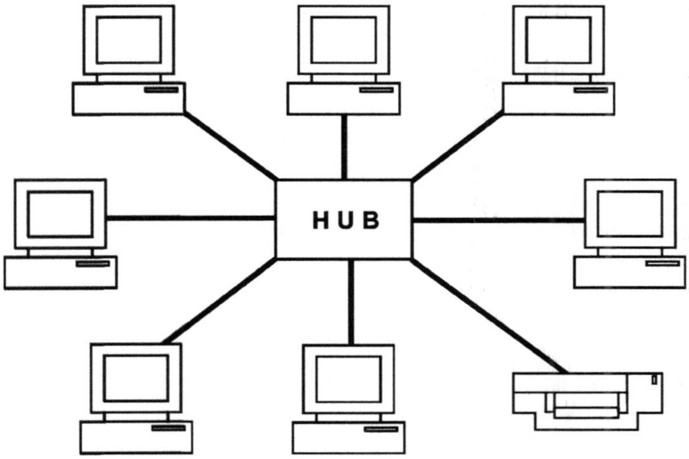

Fig. 5.22 Star topology with a hub

however, share the bandwidth, for instance, a bandwidth of 100 Mbit/s with 100 nodes means that each node has 1 Mbit/s, which is reduced still further due to collisions.

Switches

Like a bridge, a switch is used to increase the capability of an Ethernet LAN by dividing the network into several collision domains (Fig. 5.23).

Fig. 5.23 Comparing a hub (left) and a switch (right)

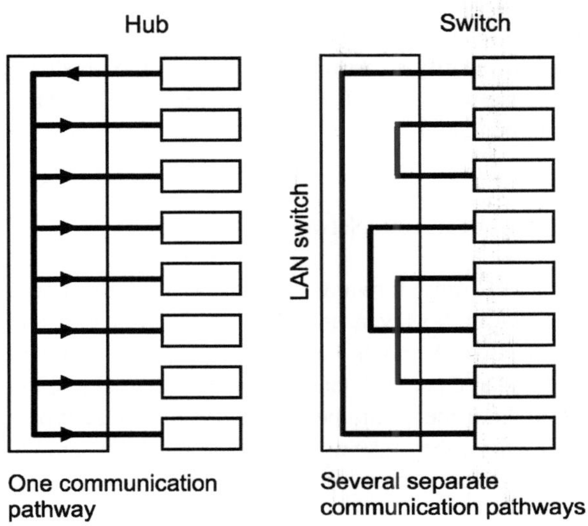

This reduces the data traffic in each network segment and increases the usable bandwidth.

A switch is a multiport bridge and is a layer two device. It receives incoming frames at its ports and forwards them to the corresponding channels. In contrast to a hub, a switch forwards the frames to specific addresses. It checks its address table to see where a frame needs to be sent. As with a bridge, the switch's address table is automatically completed over time.

In an ideal scenario, using a switch with 10 ports and 100 Mbit/s per port, it should be possible for five pairs of stations to exchange data simultaneously. The total transmission rate would then be 100 Mbit/s $\times$ 5 = 500 Mbit/s, five times that of the hub's transmission rate. If you connect a workstation to one of the switch's ports instead of another hub, you could then convert the operating mode in this microsegment from half to full duplex, effectively doubling the transmission rate again. In our example the switch would have to have an internal transmission rate of at least 1 Gbit/s to ensure that no frames are lost when at full capacity.

In practice, data traffic is hardly ever evenly distributed across all the switch's ports. On the whole, there are a large number of workstation computers that access one or more servers. The workstation computers have ports with 100 Mbit/s, whereas the servers have a 1 Gbit/s connection to the switch.

A switch's ports are generally downwardly compatible, ensuring that even older or slower components can be used. The transmission rate and operating mode (full or half duplex) is chosen automatically (auto-negotiation). The transfer rate used in building automation does not have to be that high, so even 10-Mbit/s Ethernet is normally sufficient. IT applications with different operating modes, voice-over internet protocol (VoIP) or even transferring videos can mean that a LAN reaches its capacity very quickly. Network management, however, including monitoring components and their loads, can ensure that potential bottle necks are recognized early enough. As Ethernet is a non-deterministic protocol—in which messages are sent within a specific interval of time and the data frames sent by network components may be discarded when the load is excessive—sporadically occurring and hard-to-identify errors are not unheard of. You should therefore always calculate in a reserve when installing an Ethernet LAN to take into account any unforeseen eventualities.

Virtual Local Area Networks

A switch only forwards a frame out of the port the destination station is on. Broadcasts, however, are addressed to all stations and are therefore forwarded out of all ports. A switch-based network is also known as a broadcast domain.

In a large network that has many nodes, a large number of broadcasts can be sent throughout the whole network. Because these broadcasts have to be processed by all stations, they take up part of the available bandwidth and burden the connected stations. Furthermore, in a switch network, each station can, in principle, exchange messages with every other station. For security reasons, you want to prevent this happening and only allow certain user groups to communicate with each other.

As an example, say we have three user groups at a university: professors, students and administration. These user groups are to be kept separate to prevent unauthorized access to exam questions or personal data. The usual security measures such as password protection are often insufficient or can be easily bypassed. One solution would involve creating three physically separate networks, as shown in Fig. 5.24.

Each group has a switch and the necessary cable connections. When organizational changes need to be made (e.g., an employee moves to a different office), a new cable has to be laid to connect the workstation to the correct switch. A quicker and cheaper way to separate the groups is to use virtual LANs (VLANs). With a VLAN-compatible switch, each port can be assigned to a group using the appropriate software. This means that if you need to assign a port to a different group, you do not have to lay a new cable.

Figure 5.25 shows the ports and the user groups they have been assigned to. The switch only forwards data frames between ports that belong to the same

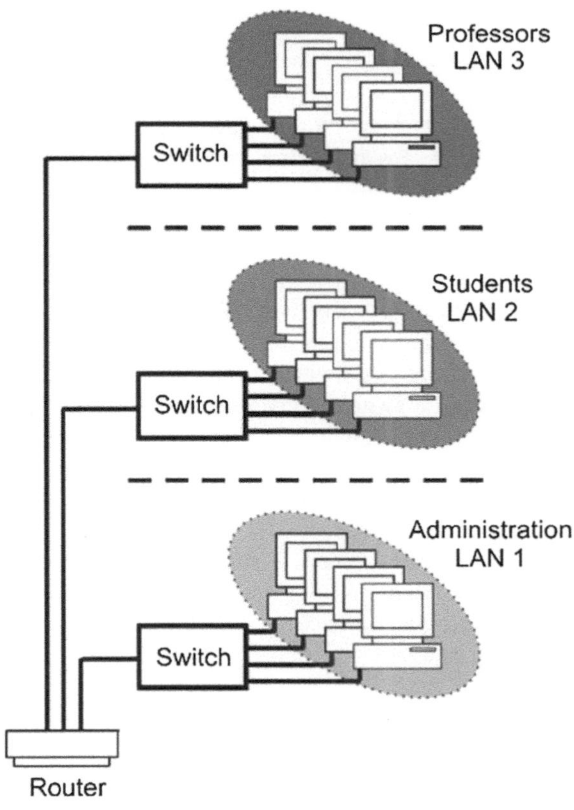

Fig. 5.24 Non-VLAN networks with three separate user groups (professors, students, administration)

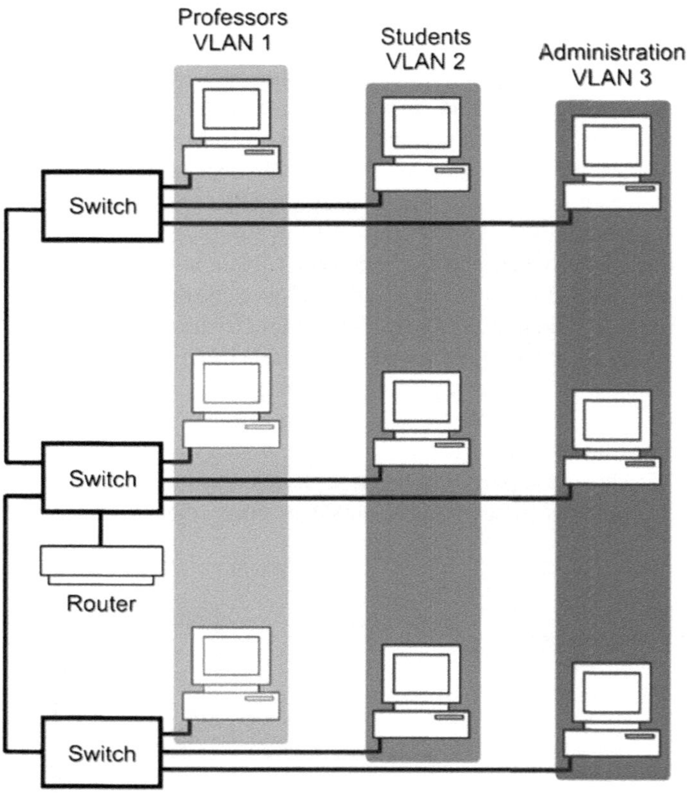

Fig. 5.25 VLAN-networks and three separate user groups (professors, students, administration)

group. Even broadcasts are only sent to computers that belong to the same group. For this method to work over a number of switches, each frame that is to be forwarded from one switch to another must be tagged—known as tagging. The frame sent to a switch is given an additional group ID. If it is a broadcast frame, then it is sent to all switches. The switches then remove the group ID and forward the frame to all computers that belong to this group. For the computers this process is completely transparent because the changes made to the frame by the transmitting switch can be reversed.

If, however, you want to allow communication between different groups, you will need a router. A router can forward data to different groups and a firewall prevents unauthorized access.

5.2.3.3 Transmission Over Fiber-Optic Cable

Light pulses are used to send data over a fiber-optic cable. This technology is more expensive than using electrical communication cable, but still has the following advantages:

- Transmission:
  - Low signal attenuation
  - High bandwidth and therefore a high transfer rate
  - More secure than electric communication cable or radio communication
- Interference:
  - Not affected by lightning or high-voltage/power lines
  - Can be used in potentially explosive areas (no sparks)
  - No signal radiation (high electromagnetic compatibility)
  - Network components are galvanically separated (no equalizing current due to potential difference)
- Physical:
  - Small diameter and weighs little (save space, easy to lay)
  - Large supply length and long cable length without the need for intermediate stations
  - Not susceptible to corrosion

The disadvantages are the high cost and the complex process of fitting connectors. Furthermore, fiber-optic cable must be laid as straight as possible to prevent damage to the waveguide.

The Structure of a Fiber-Optic Cable

A fiber-optic cable consists of an inner and outer core surrounded by a cladding layer, all of which are made of fused quartz (silicon dioxide) (Fig. 5.26). The core

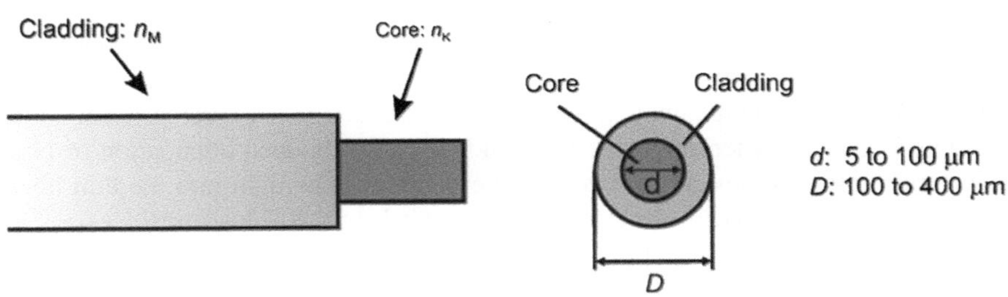

Fig. 5.26 Structure of an optical fiber

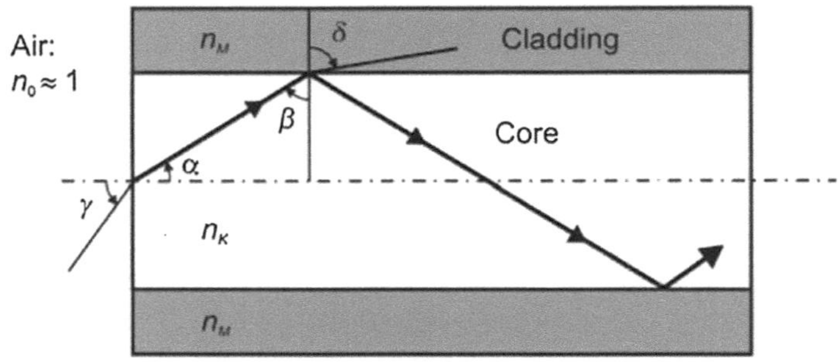

Fig. 5.27 Injection and propagation of light in a fiber-optic cable

and the cladding are differently “doped” (intentional introduction of impurities) to create the desired optical refraction index.

According to the laws of optics, this means that a light ray entering the fiber will be reflected at the core-cladding boundary, as long as the core’s refraction index is greater than that of the cladding. Repeated total internal reflection allows the light ray to travel along the fiber-optic cable (Fig. 5.27). You can use the laws of refraction to determine the angle.

To protect the sensitive, hair-thin optical fibers from physical damage, the fibers can be enclosed in a protective plastic layer (e.g., Kevlar). Fibers can also be made of plastic instead of fused quartz. Due to their poorer optical attributes, these plastic fibers are not used for high-speed data transmission over long distances, but only for digital audio transmission (connecting a CD/DVD player to an amplifier. Polymer optical fiber is also used for connecting control devices in automobiles. Research is also being carried out into developing cheap high-speed systems to replace DSL.

Attenuation

Attenuation is the loss of light signal strength as the signal travels from the transmitter to the receiver. This should be as low as possible to ensure that the signal can travel as far as possible.

Attenuation in fiber-optic cable can have many causes. Intrinsic loss is caused by Rayleigh scattering and infrared absorption; whereas extrinsic loss is caused by material impurities, which also cause light to scatter, therefore impeding the transmission of the light signal along the fiber. Attenuation also increases the more the fiber is bent or when the cable is extended (splice-induced attenuation or plug connection loss). Losses can also occur when directing the light into the thin fiber. Attenuation is measured in decibels (dB) per unit length (dB/km).

Attenuation A in dB can be calculated as:

$$A = 10 \times \log \frac{P_S}{P_D}$$

P_s is the signal power at the transmitting end (source) and P_D is the signal power at the receiving end (destination).

Typical attenuation values for fiber-optic cable are between 0.5 dB/km and 3 dB/km. As attenuation in optical fibers is dependent on frequency, an optical window of 850, 1300, and 1550 nm is used, within which attenuation is particularly low.

Dispersion

The light guided into a fiber-optic cable can travel along the length of the cable when reflected at a particular angle (Fig. 5.28). This happens with so-called multimode fiber, which has a comparatively large core diameter of 50 μm.

Rays of light that enter the optical fiber at different angles have different path lengths and therefore arrive at the receiving end at different times. This leads to modal dispersion. When using light to transmit signals, you must therefore make sure that the gap between each of the bits is large enough; otherwise they will "pile up" at the receiving end, making it difficult or even impossible to recognize the signal's logical state (0 or 1).

A fiber-optic transmission system is often characterized by its bandwidth-distance product. This is because the effect of dispersion increases with the length of the fiber. The bandwidth-distance product gives the maximum possible transmission speed over a length of fiber-optic cable. For example, a fiber with a bandwidth-distance product of 1000 MHz km can transfer a 1000 MHz signal over a distance of 1 km (transmission rate of 1000 Mbit/s) or a 500 MHz signal over a distance of 2 km (500 Mbit/s).

Modal dispersion can be prevented by reducing the core's diameter. If the diameter of the core is only a few micrometers (μm), the light in the waveguide propagates in almost a straight line. These fibers are known as single-mode fibers because they have only one propagation path and display no modal dispersion. Single-mode fibers therefore have a higher bandwidth-distance product compared to multimode fibers (100 GHz km instead of 1 GHz km).

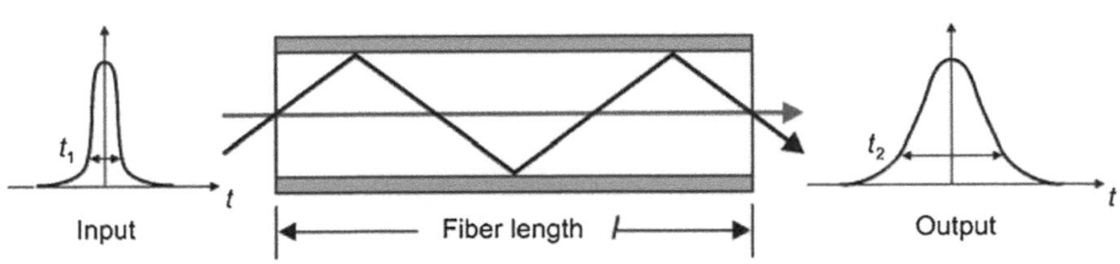

Fig. 5.28 Propagation paths in a multimode optical fiber

The maximum transmission rate of a single-mode fiber is limited by material dispersion. The components of light have different wavelengths (colors) and these separate at different velocities in the optical fiber. A light pulse with a particular spectral width splits into its wavelength components and "fans out". This effect can only be reduced by using light sources that have particularly high color purity such as lasers.

Optical Transmission Systems

As well as fiber-optic cable, an optical transmission system also consists of the components shown in Fig. 5.29.

Semiconductor lasers or light-emitting diodes (LEDs) are used as transmitters. The electrical data signal controls the light intensity of the transmitter. The lasers are particularly suitable as fiber-optic transmitters because they have a narrow spectral bandwidth (low material dispersion) and concentration of rays of light, which means that the light pulses can be easily funneled into the fiber. A photodiode is used as a receiver to convert the light pulses back into an electrical signal. Repeaters may have to be used over longer stretches of fiber-optic cable. The repeater converts the light signal into an electrical signal, regenerates it (PR—pulse regeneration) and then converts it back into a light signal. Over extremely long distances optical receivers are often used. In the future, we should see the widespread use of other optical components (switches, filters, couplers).

Types of Fiber-Optic Cable

Table 5.3 shows the preferred types of fiber-optic transmission for Ethernet. The 1-Gbit/s standard is the most widely used.

Separate cables are used to transfer data to and from the receiver (Tx or Rx). These separate signal paths for Tx and Rx facilitate full-duplex. Straight Tip (ST) connectors and Standard Connectors (SC) are used.

For 1000Base-SX multimode fiber is used with a wavelength of 850 nm. All the components must have the same diameter, 50 and 62.5 μm are standard. 1000Base-LX with single-mode fiber (core diameter of 10 μm) can cover the

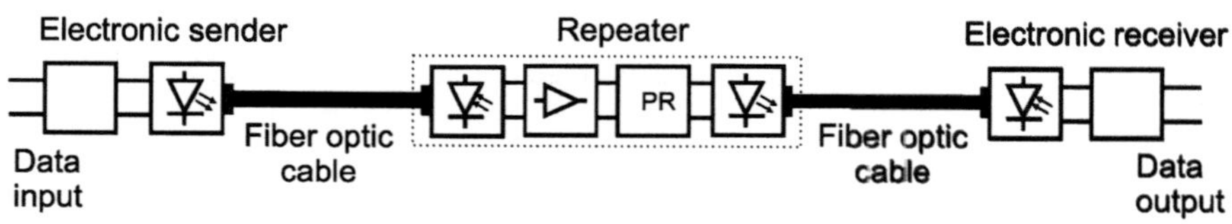

Fig. 5.29 An optical transmission system

Table 5.3 Ethernet versions of fiber-optic transmission

Name	Speed	Maximum fiber length
100Base-FX	100 Mbit/s	~400 m
1000Base-SX	1 Gbit/s	~500 m
1000Base-LX	1 Gbit/s	~5 km

furthest distances, up to 5 km, and uses a wavelength of 1300 nm. More powerful 10-Gbit/s Ethernet standards have also been developed (10Gbase-LX4, 10Gbase-SR, 10Gbase-LR, and so on), which vary depending on the fiber used, wavelength range and maximum distance.

5.2.3.4 Structured Cabling

The aim of structured cabling (Fig. 5.30) is to use as few types of media and networks as possible to transmit as many applications as possible.

The European standard EN 50713 defines criteria for manufacturer and application-neutral cabling for flexible and long-lasting cabling. There are three levels for three different areas of use.

- The primary area deals with cabling between buildings, beginning and ending at the main distributor in the buildings. Fiber-optic cable is used here due to the relatively large distance and to prevent equalizing currents caused by differences in potential.
- The secondary area refers to the backbone within a building that connects the main building distributor with each of the floor distributors. Both copper and fiber-optic cable can be used here.
- The tertiary area is where all the terminals are connected to the floor distributor, usually with twisted-pair cable in a star topology. Depending on the conditions on site, instead of a floor distributor you can connect all the terminals in a star formation to a powerful network component, which is then connected to the building distributor.

With current trends such as voice-over IP (VoIP), structured cabling can be used for establishing a network of computers as well as for making telephone calls. Using structured cabling in building automation seems then only logical. However, you should not ignore potential security issues. Building automation systems could

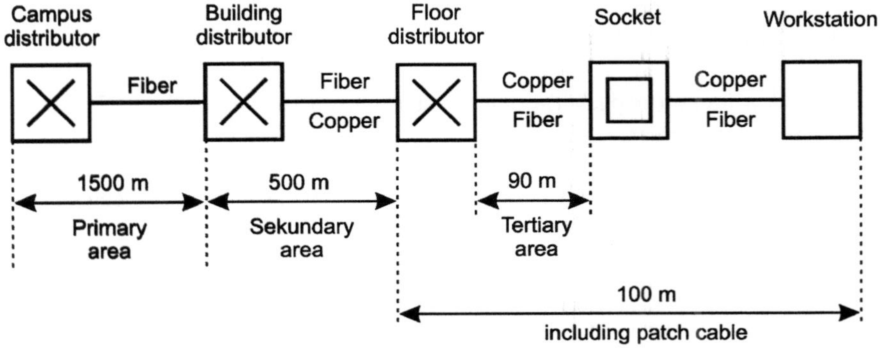

Fig. 5.30 Structured cabling

be accessed from every office computer relatively easily. Moreover, you must also ensure that reliable building automation functions are not jeopardized by additional services overloading the network.

5.2.3.5 Wireless LAN

Most Ethernet networks used in building automation use coaxial, twisted-pair or fiber-optic cable. Wireless LAN (WLAN) offers an alternative that can be used to connect remote buildings or installed in landmark-protected old buildings, in which laying new cable is prohibited.

Transmission with WLAN is transparent. This means that you will not normally notice the difference between WLAN and cable transmission. WLAN, however, is less reliable and more prone to interference and eavesdropping. Table 5.4 shows three standards from the Institute of Electrical and Electronics Engineers (IEEE).

The actual transmission rates are about half that of the speeds shown in the table above. This is due to half duplex transmission and the control overhead needed to regulate shared access to the radio channel.

The most widespread standards are 802.11b and 802.11 g, which use the 2.4 GHz frequency band. The radio channels available are, to a certain extent, already oversubscribed. Other wireless technologies such as Bluetooth, Zigbee and wireless audio/transmission also use the same frequency band, which can lead to mutual interference. Because you do not need a license or to register in order to use these technologies, there is no guarantee of protection against interference. The limited range (about 300 m outdoors) means that systems that use the same frequency can be isolated. At present, the 5 GHz frequency allocation is considerably lower. The drawback with higher frequencies, however, is that they have a far lower range in buildings compared to lower frequencies. This is because walls impede the signals. Planning WLAN systems is, therefore, relatively difficult and requires that measurements be frequently taken on site to determine the range or area of coverage.

5.2.3.6 Frames and MAC Addresses

Ethernet is a frame (or packet)-oriented data transmission technology. The transmitter breaks up the user data, embeds it into frames and then transmits the frames over the transport medium to the receiver. The receiver then extracts the user data. The advantage of this method is that a number of connected stations can use the

Table 5.4 WLAN standards

Name	Speed (Mbit/s)	Frequency range (GHz)
802.11b	Up to 11	2.4
802.11g	Up to 54	2.4
802.11a	Up to 54	5

available medium seemingly simultaneously. In reality the frames from each node are sent one after another. A station (node) can access the transmission medium again after a delay of up to 1518 bytes (max. length of a frame used with cable-based Ethernet).

Figure 5.31 Shows the Structure of an Ethernet Frame.

The preamble tells the connected stations that the transmission is about to begin. This is followed by the recipient's hardware address, which tells the station whether the message is addressed to it or whether it can be ignored. The IEEE allocates the six-byte (48 bit) media access control (MAC) address (Fig. 5.32) to senders and receivers.

The individual/group (I/G) bit shows whether the message is addressed to one station (unicast) or a group of stations (multicast). The universal/local (U/L) bit shows whether the MAC address conforms to IEEE conventions or whether it has been assigned by a local network operator. There are 22 bits available for the vendor ID assigned by the manufacturer, followed by 24 bits for the device or serial number.

You can change the MAC address of most network cards using a specific software tool. You should always make sure, however, that you do not assign the same address to more than one node in a local network. The 0xFFFFFFFFFFFF MAC address is of particular importance, because it is a broadcast address, in other words, it is sent to all stations within a network.

The sender's MAC address is followed by the most commonly used Ethernet frame format, a two-byte type field used for identifying the user data. 0×0800 indicates, for example, that an IP packet is in the user data area. The user data itself can be between 46 and 1500 bytes long. It is followed by a 4 byte frame check sequence for detecting errors.

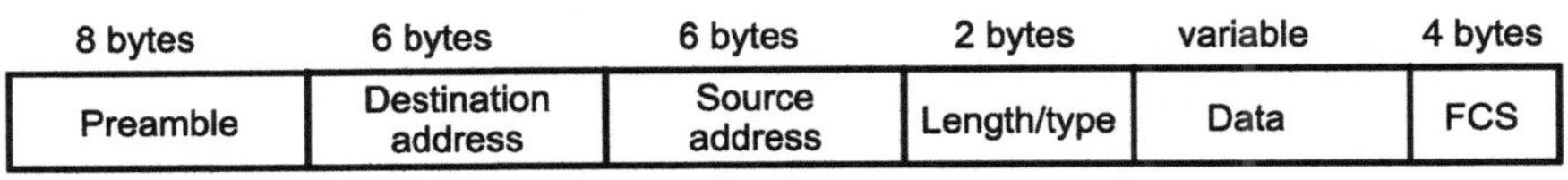

Fig. 5.31 Structure of an Ethernet frame

Fig. 5.32 The structure of a MAC address

5.2.4 Arcnet

Attached Resource Computer Network (ARCNET), standardized in ATA/ANSI 878.1, is a local area network (LAN) technology developed in the USA for office intranets, but which has now been superseded by the faster Ethernet. Unlike Ethernet, however, ARCNET is deterministic, which makes it particularly good for use in industrial communication technology. Like MS/TP, ARCNET incorporates token-passing, which has the following benefits:

- It functions in real-time, that is, able to accurately predict the time it takes to answer
- OSI layer 1 and 2 implemented in hardware (chip)
- Variable topology (bus, tree, star)
- A variety of media options (coaxial, twisted-pair, fiber-optic cable)
- Variable bit rates from 30bit/s to 10 Mbit/s
- High efficiency through low overhead (transmission rate up to 71% of the Baud rate).

ARCNET is not, however, that widely used outside of the USA, which means that only a few BACnet devices actually support it.

5.2.5 LonTalk

LonTalk is another LAN technology that can be used to transport BACnet messages. It is a communication protocol developed by the company Echelon as part of the family of LonWorks protocols, described in detail in Chap. 4. Transmission media used include twisted-pair, coaxial, and fiber-optic cable as well as radio systems that allow scalable transmission speeds of up to 1.25 Mbit/s. Being able to send BACnet messages over LonTalk does not mean, however, that the higher levels of BACnet and LON can communicate with each other. LonTalk is just the transport medium; gateways connect the BACnet objects with LonWorks/LonMark.

5.3 The Network Layer

5.3.1 Purpose

The purpose of the network layer is to relay messages between different networks, regardless of the layer-two technology used in these networks (Fig. 5.33).

The data link layer only deals with local addresses, whereas the network layer must enable global addressing. For example, there are 254 local addresses and one

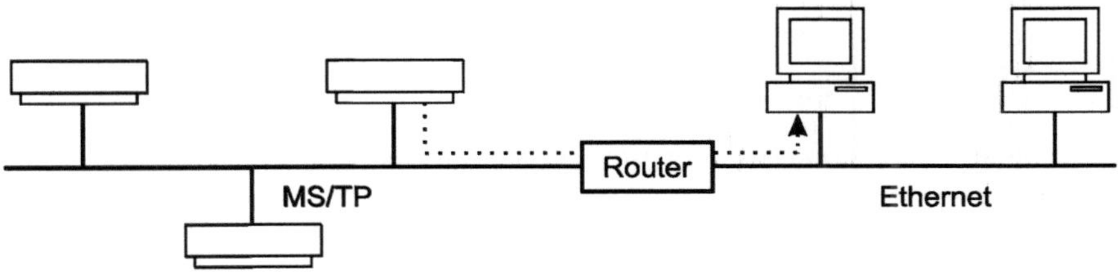

Fig. 5.33 Two networks connected over a router

broadcast address available in an MS/TP network. A second MS/TP network has the same number of potential addresses. If you want to connect the two networks, you need an additional way of identifying the stations within the whole network. In BACnet this is achieved using a two-byte network number (maximum of 65,535 subnetworks). A BACnet station is therefore uniquely identified by the network number and the layer-two address.

5.3.2 Routers

A router is a device that connects two or more networks and decides whether messages should be forwarded. A router can connect networks of the same or different layer-two technology. This functionality can also be integrated into a BACnet automation station. A router uses a routing table to decide whether to forward a message to another network. These routing tables can be created automatically by the exchange of path information between routers. A so-called routing protocol defines how the information is exchanged. Unlike the Internet, the method used in BACnet is very simple. The networks, however, must be configured so that there is only one active pathway between two stations at any one time.

The network layer inserts a header before the application layer protocol data unit (APDU) (Fig. 5.34).

The first octet of the network layer protocol data unit (NPDU) contains the version number of the BACnet protocol (currently "1"). This is followed by a control field, which indicates whether it is a routing message (routing protocol message) or a BACnet message (data from the BACnet application). In addition to the addresses, the length of the MAC addresses must also be supplied—ranging from one byte (MS/TP) to seven bytes (LonTalk with a Neuron ID).

A message sent from a BACnet station to a router contains the destination address. The source address is determined from the local address at the data link layer. Whichever one of the router's ports the message enters shows which network the message has been sent from. NPDUs sent between routers contain source and destination addresses, whereas the messages from a router to a BACnet device contain only the source address.

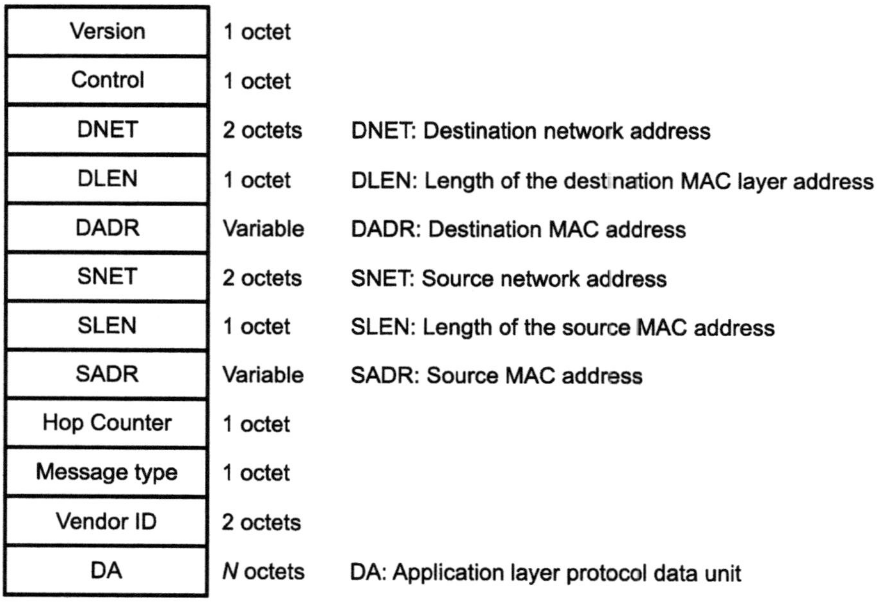

Fig. 5.34 The structure of a network layer protocol data unit (NPDU)

The hop count is a decrementing counter value that ensures that BACnet messages cannot be routed in a loop indefinitely. This only occurs if the configuration rule that allows only a single path between any two BACnet nodes is not adhered to. Each router the message passes through reduces the hop count by one. If the hop count reaches zero, the message is discarded.

Depending on the task, certain fields in the header may not be needed and are omitted. For example, only the version and control fields are used in BACnet messages exchanged between two stations in a local network (Fig. 5.35).

The exchange of routing messages has also been standardized. Below are a few of the standardized message types:

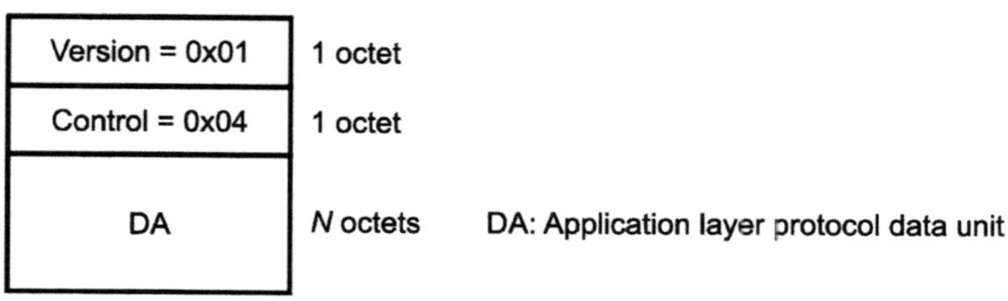

Fig. 5.35 NPDU transmitted between two stations in the same network

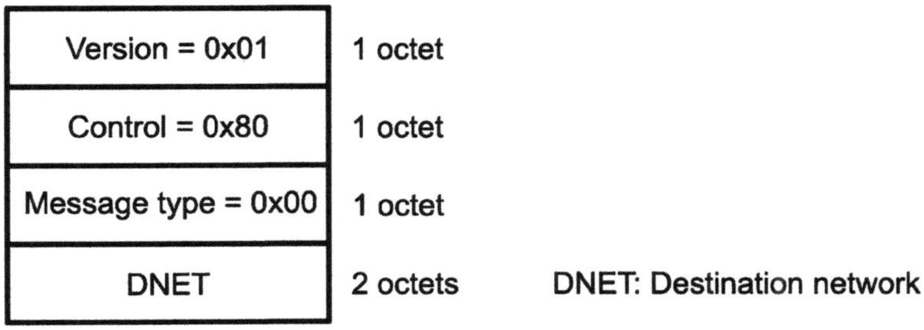

Fig. 5.36 Example of a Who-Is-Router-To-Network message

- Who-Is-Router-To-Network
- I-Am-Router-To-Network (frame format see Fig. 5.36)
- I-Could-Be-Router-To-Network
- Reject-Message-To-Network
- Initialize-Routing-Table.

5.3.3 BACnet and Internet Protocols

The use and importance of the Internet worldwide has resulted in its protocols being increasingly adopted in building automation. BACnet supports Internet Protocol (IP) and therefore enables global networks to be interconnected. IP is a network layer (OSI layer three) protocol and is responsible for routing packets through a complex network. This requires the source and destination addresses as well as best-possible path through the network.

5.3.3.1 IP Addresses

IP addressing has a hierarchical structure, unlike the flat structure used in Ethernet. In a flat addressing system, the address does not contain any information as to the location of the station and any paths to it. This is not that necessary in a medium-sized local network. If, however, you were to use a flat addressing system in the Internet, you would have to store the pathways for each individual station in the router, which is not only extremely inefficient but is also technically almost impossible.

An example of an addressing system used for large networks is the hierarchically organized telephone networks. An area code is used for identifying the local telephone network, followed by a telephone number for the subscriber within this network. IP addresses are organized in a similar way. They consist of 32-bit

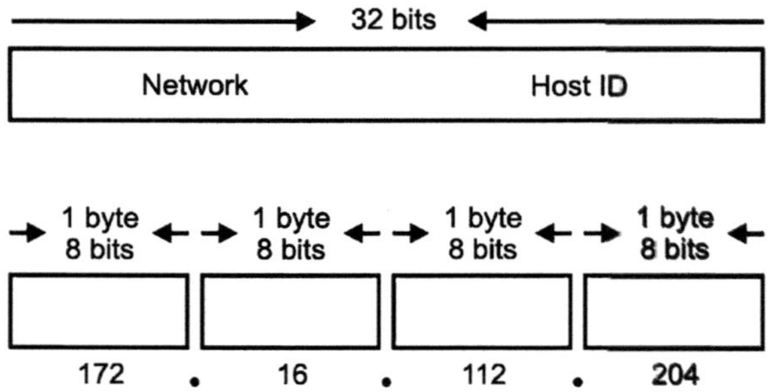

Fig. 5.37 Format of an IP address

Table 5.5 The five classes of IP addresses

Class	Leading bits	Value of the first byte	Bits for the network number	Bits for the host number	Maximum number of hosts
A	0	1–126	8	24	~16 million
B	10	128–191	16	16	~65,000
C	110	192–223	24	8	~250
D	1110	224–239	Multicast		
E	1111	Reserved for experimental purposes			

number, usually represented by four octets (Fig. 5.37). The decimal format makes the number easier to read for the human eye.

The Internet Assigned Numbers Authority (IANA) is responsible for allocating globally unique IP addresses.

An IP address is divided into two parts: the network number and the host number. The network number is equivalent to a telephone number's area code; whereas the host number corresponds to the telephone number itself within a local telephone network. The number of the available 32 bits used for the network and host numbers varies. IP addressing supports five different address classes, each of which has a different network-to-host bit ratio (see Table 5.5).

Class A

Class A addresses are used for very large networks with many hosts. As a result, there are more bits available for the host number and less for the network number—the first octet is for the network number and the other three are for the host number. The first octet can only contain values between 1 and 126, which means that there can only be 126 class-A networks, each with $2^{24}-2 = 16{,}777{,}214$ hosts. From the possible 2^{24} combinations in the host portion, one address is reserved for the

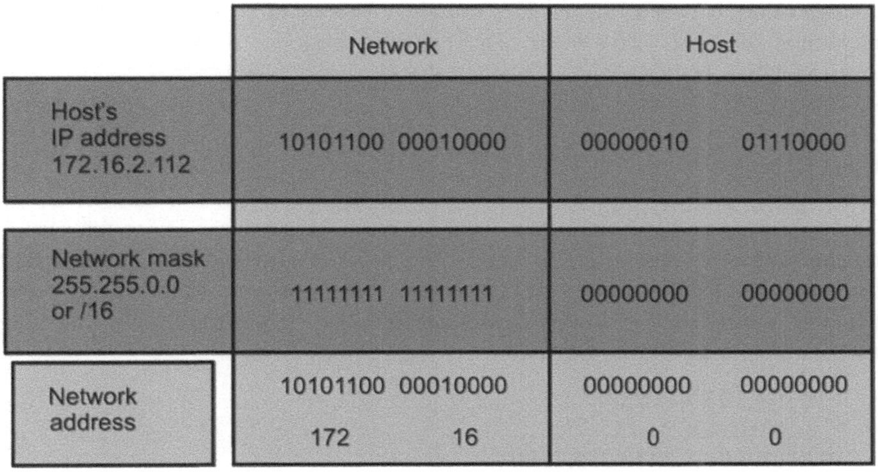

Fig. 5.38 Logical AND operator for creating network addresses from IP addresses and subnet masks

network (all bits in the host part are then "0") and one bit for the broadcast address (all bits in the host segment are then "1").

To identify the network and host sections, another 32-bit number is introduced as a so-called network mask. This is also usually represented as four octets, written in decimal form and separated by periods. In binary form the subnet mask has binary 1 s in all bits specifying the network and field, and binary 0 s in all bits specifying the host field (Fig. 5.38).

You can quickly identify which network address the IP address belongs to using a logical AND operation of IP addresses and network mask. The network mask for a class-A network is 255.0.0.0.

Class B

Class B addresses are reserved for medium-large to large networks. The network and host portions are each 16 bits in length. The first octet must only contain values between 128 and 191 (the leading two bits are 10). This results in a total of 2^{14} networks, which are allowed to contain up to $2^{16}-2 = 65{,}534$ hosts. The network mask for a class-B network is 255.255.0.0.

Class C

Class C addresses are used for smaller networks. The first three octets are reserved for the network number and the last octet is for the host number. The first octet can contain values between 192 and 223, resulting in a total of 2^{21} networks each with a maximum of $2^8-2 = 254$ hosts. The network mask for a class C network is 255.255.255.0.

Other Classes

The remaining address ranges are used for multicasting (class D) or for experimental purposes (class E). The address 127.0.0.1 is also known as a local host address. This address is used to contact the actual host computer, which is particularly beneficial for testing and checking server services. The following addresses are only for private use:

- 10.0.0.0–10.255.255.255
- 172.16.0.0–172.31.255.255
- 192.168.0.0–192.168.255.255

These are private addresses that can be used arbitrarily in local networks, but should not end up on the Internet or forwarded by a router.

Due to the lack of available public IP addresses, private IP addresses are often used. A router with network address translation (NAT) capabilities connects the network to the Internet. The router converts the local IP addresses into one public address, which represents the whole private network on the Internet.

5.3.3.2 Routing

Routers use routing protocols to share path information with other routers. This information is used to build tables that store the routes (paths) to particular network destinations.

As well as dynamic routing, an administrator can also enter information in the routing table manually (static routing). This has the advantage that it reduces the load on the network connections that normally occurs in dynamic routing when routers exchange routing information. The drawback, however, is that if a path fails, a replacement path is not automatically found, unlike with dynamic routing.

Figure 5.39 shows a network with three routers. A local network with one computer is connected to each of these routers.

The local networks have the network addresses 10.0.0.0, 11.0.0.0, and 12.0.0.0. The paths between the routers are also networks and are therefore also assigned network addresses (13.0.0.0, 14.0.0.0, and 15.0.0.0). Each port on a router has its own IP address. The ports are also marked with either E0 for Ethernet or S0 or S1 for the serial connections. The routing tables are constructed like the one shown in Table 5.6.

When computer 1 sends a packet (frame) to computer 2, it arrives first at router 1. This router determines the destination network (12.0.0.0) from the destination IP

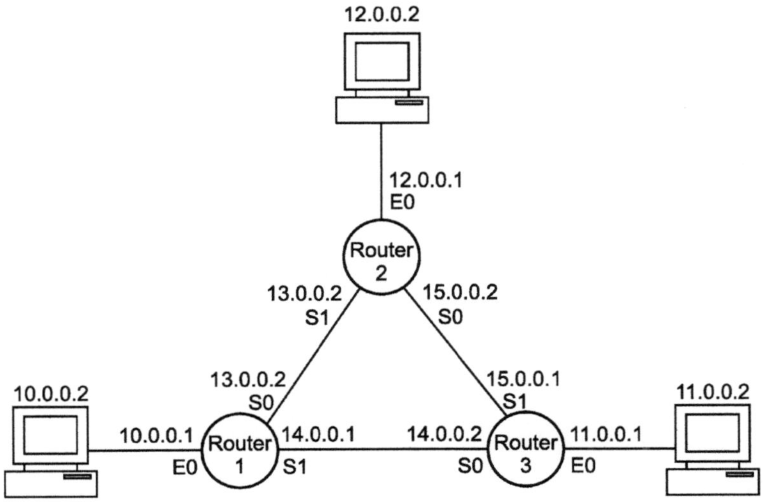

Fig. 5.39 Example of a network with three routers

Table 5.6 Example of a routing table for the network shown in Fig. 5.39

Network	Entries for the routing ports		
	Router 1	Router 2	Router 3
10.0.0.0	E0	S1	S0
11.0.0.0	S1	S0	E0
12.0.0.0	S0	E0	S1
13.0.0.0	S0	S1	S1
14.0.0.0	S1	S1	S0
15.0.0.0	S1	S0	S1

address (12.0.0.01) using the subnet mask (255.0.0.0—class A) and then forwards the packet through the appropriate port (S0). In same way, the next router (router 2) determines the path to computer 2. The IP packet takes two routers or hops to reach its destination.

5.3.3.3 Packets

An IP packet consists of an IP header followed by data. The header is normally 20 bytes long. The fields are displayed in four-byte blocks, one under the other, instead of in a row (Fig. 5.40).

The main entries in the header are:

- IP version (currently version 4)
- Source and destination IP addresses

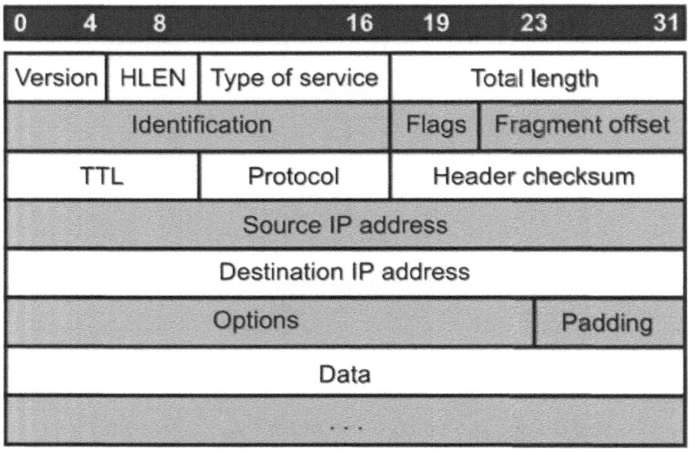

Fig. 5.40 Contents of an IP header

- Protocol (shows which OSI transport protocol follows)
- Checksum for the header
- Time to live (TTL)

The TTL field contains a value that is decremented by 1 every time the packet is forwarded by a router (the same as the hop count in BACnet). When the value is zero, the packet is deemed as invalid and is discarded. This prevents an IP packet from traveling in an endless loop (due to a potential error in the routing table), creating unnecessary data traffic. The lifetime of an IP packet is therefore limited. Errors in the routing tables can arise when the paths have to be recalculated after a section failure. If this happens, it can take several minutes before the network is reconfigured and all routing tables are up to date again.

With the help of the TTL field you can also track the path taken by packets across a network. To do this, the Internet Control Message Protocol (ICMP) is used; whose best-known function is the echo.

A ping command generates and then sends an ICMP "echo request" packet from most operating systems. The destination computer replies by sending back an ICMP echo reply, as long as the service has been activated or the packet has not been blocked by a firewall. This enables you to test whether a computer can be reached and measure the round-trip time of an IP packet.

By sending ICMP echo requests with different TTL values in the IP header, you can determine which routers are on the path to the destination computer. You can automatically trace the route taken by a packet using a trace request such as tracert (Windows) or `traceroute` (Linux). This way you can benefit from error messages created by routers when the TTL value of a packet or the number of allowed hops has been exceeded. The trace request successively increases the TTL value of each packet sent and analyzes the error messages from the routers on route to the destination. Thus the IP packet with TTL value "1" is discarded by router one, which then returns an error message; packet with TTL value "2" is discarded by router two, and so on.

5.3.3.4 Subnetworks

Due to the explosive growth of the Internet, the remaining available IP addresses have to be used sparingly. This is achieved using subnetworks (subnets). Figure 5.41 shows you an efficient way to assign IP addresses.

Two sites, each with 100 computers, are to be connected with each other over a router. Without subnets, you would have to assign each site a class C network with 254 possible computers. You would therefore take up 508 IP addresses, of which only 308 would actually be used.

A more efficient solution is to divide one class C network into two subnets. The IP addresses are then split into three parts: the network, subnet and host fields. The subnet field is created by borrowing bits from the host field.

Example: the class B network 172.16.0.0 is to be divided into subnets (Fig. 5.42).

Eight bits are used for subnet addressing. The previously 16-bit host field is now only eight bits long. Eight bits means 256 different subnets are available, of which, however, only 254 are actually used. The first and the last subnet are normally not used.

Each subnet has an eight-bit host field, which corresponds to 254 computers (hosts) (2^8–2). If all host bits are either "0" or "1," these are reserved as network and broadcast addresses. In the case, the network mask is 255.255.255.0, because the position of a binary "1" refers to either a network or a host field. Alternatively, instead of the network mask, you could use the IP address with the number of network mask bits separated by a slash at the end, for example, 172.16.2.0/24 is equivalent to 24 bits for the network and subnet fields.

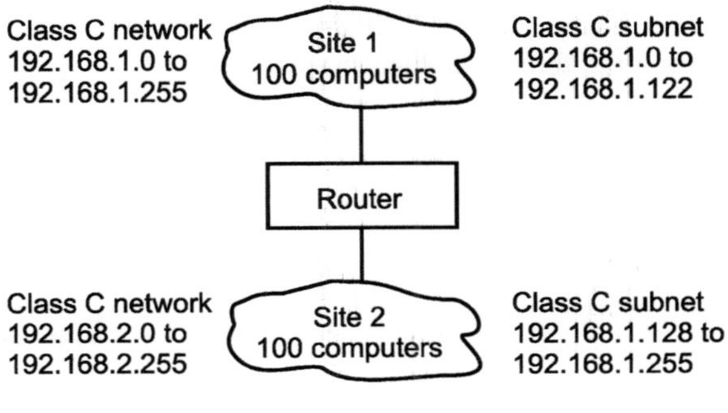

Fig. 5.41 Connecting two sites without subnets (left) and with subnets (right)

	Network	Subnet	Host ID
Host's IP address 172.16.2.112	10101100 00010000	00000010	01110000
Subnet mask 255.255.255.0 or /24	11111111 11111111	11111111	00000000
Subnet	10101100 00010000 172 16	00000010 2	01110000 0

Fig. 5.42 Class B network divided into subnets

5.3.4 Transmission Control Protocol

Above the network layer in the OSI model is the transport layer. One of the transport layer protocols is the Transmission Control Protocol (TCP). Instead of TCP, however, BACnet uses the User Datagram Protocol (UDP), which will be explained in Sect. 5.3.5. TCP provides protocols at higher levels with reliable stream data transfer. Imagine TCP as a virtual pipe for transporting data between two terminals that should guarantee that the destination terminal receives the data in the correct, uninterrupted sequence and without loss (Fig. 5.43).

A connection is established using a "three-way handshake" mechanism, the two terminals synchronize themselves by exchanging three messages, in other words, they declare themselves ready to send and receive frames in both directions (Fig. 5.44).

TCP uses IP packet-based transmission and must therefore split the data stream from the upper-level protocol into segments, which are then forwarded to IP. IP itself does not provide a mechanism to resend data packets should they get lost or damaged. TCP achieves this by assigning each packet a sequence number, usually found in the 20-byte TCP header (Fig. 5.45). The start numbers are exchanged during the three-way handshake.

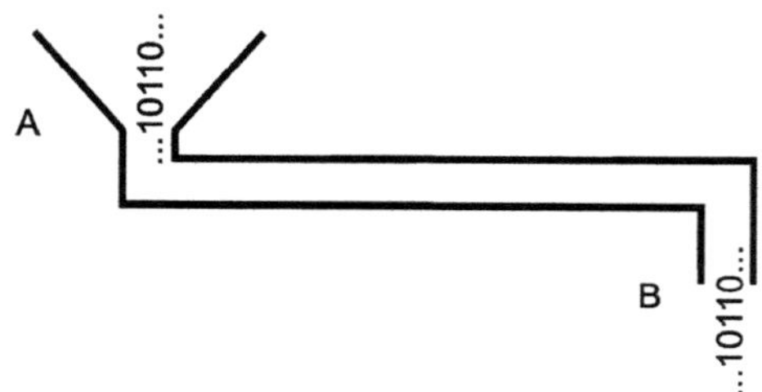

Fig. 5.43 Virtual data pipe. Bits are transported from A to B in their original sequence and without any errors

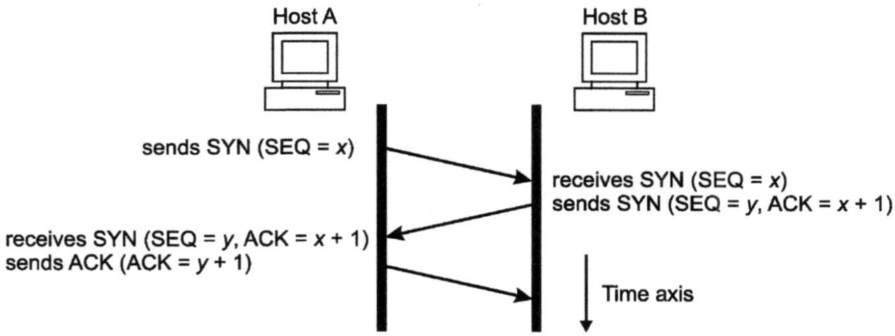

Fig. 5.44 A three-way handshake

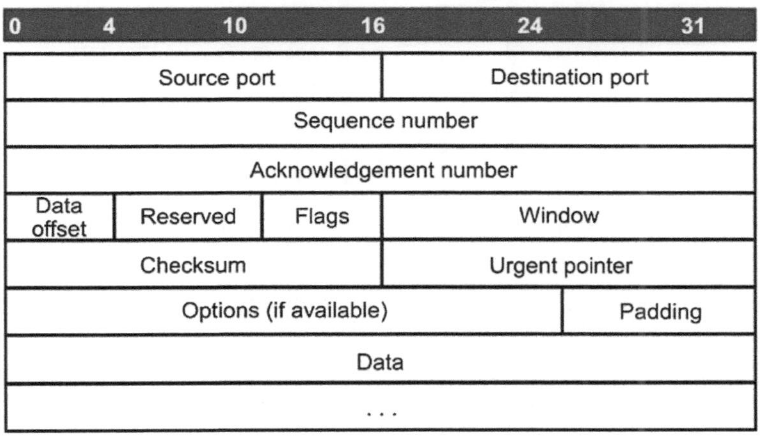

Fig. 5.45 The structure of the TCP header (each line has 32 bits)

The sequence number enables the receiver to reassemble the incoming segments in the right order. At the same time, the receiver (destination) sends back an acknowledgement number that notifies the sender (source) that has received the segment correctly. If the sender does not receive acknowledgement within a specific amount of time, or at all, it retransmits the segment. The receiver recognizes the duplicate from its sequence number, discards it and then sends another acknowledgement segment.

If the sender always has to wait until it receives confirmation of receipt before it transmits the next segment (stop and wait), this means that the throughput is very low. For every transit time between the sender and the receiver, there is a long phase during which no data is transmitted (Fig. 5.46).

For this reason, more than one segment can remain "unacknowledged" at any time, that is, allowed to be unacknowledged. During the time it takes a segment to be sent and the corresponding acknowledgement to be received, other data is

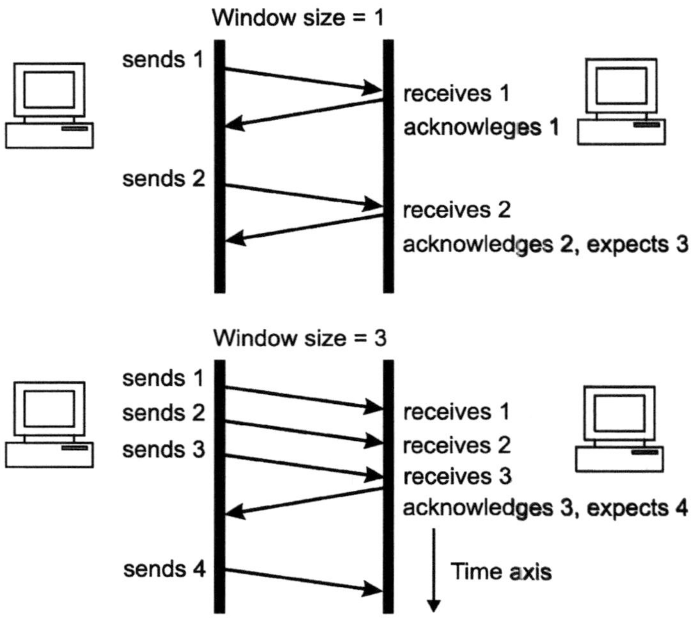

Fig. 5.46 Stop-and-wait data transmission (window size = 1) and pipelining (window size > 1)

transmitted (pipelining). The number of segments that can be sent before the acknowledgement for the first segment arrives is called the window. TCP, however, uses the number of octets as opposed to the number of segments.

Two other important fields in the TCP header are the source and destination port numbers. These numbers make sure that the TCP segments are sent to the correct application program on the destination server. For example, when you surf the Internet, receive e-mails and download a file at the same, three connections are established from your computer. Each of the programs used needs a unique port number.

The combination of an IP address and a port is called a socket. A connection between two computers can be clearly established over a pair of source and destination sockets, along with the transport protocol (TCP or UDP) (Fig. 5.47).

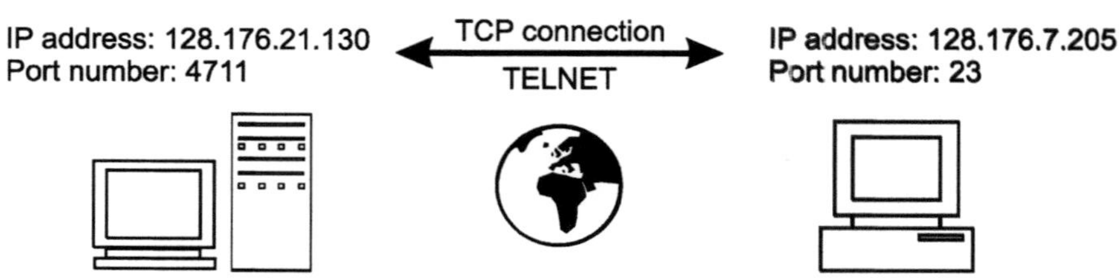

Fig. 5.47 A connection using IP addresses, port numbers and a transport protocol

The port numbers are 16 bits in length; values under 1024 are the so-called well-known ports and are reserved for specific services. For example, port number 80 is usually used to contact Web servers. The requesting client can randomly choose its port from between 1024 and 65,536. Certain manufacturers also use some of the port numbers in this range.

5.3.5 User Datagram Protocol

The way TCP establishes and terminates connections means that it is too complex for BACnet. BACnet therefore uses the User Datagram Protocol (UDP), which also belongs to the Internet Protocol family. This is a very simple protocol that is used for transporting datagrams without acknowledgement or guaranteed delivery. The BACnet application layer is responsible for controlling the retransmission of a packet should it go astray. Figure 5.48 shows the simple structure of a UDP header.

5.3.6 ARP and DHCP

A connection must be established between MAC addresses and IP addresses at the interface between layers two and three. This and the automatic assignment of IP addresses are carried out by the following two protocols.

5.3.6.1 The Address Resolution Protocol

To communicate in an Ethernet network using TCP/IP, a sender needs to know the receiver's MAC address as well as its IP address. ARP is used to automatically determine the destination host's MAC address and map it to the IP address.

Figure 5.49 shows an example of a client (computer with a Web browser, 10.1.1.3) and a Web server (10.1.1.6) in the same network. After you have entered the Web server's IP address in the Web browser's address field, the operating system first uses an Ethernet broadcast to send an ARP request asking for the

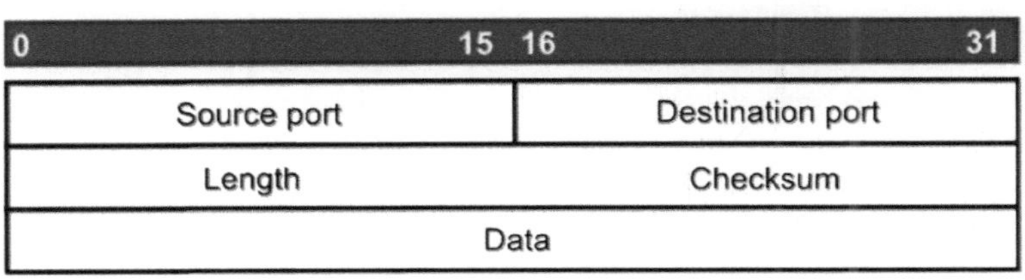

Fig. 5.48 The UDP header (each line has 32 bits)

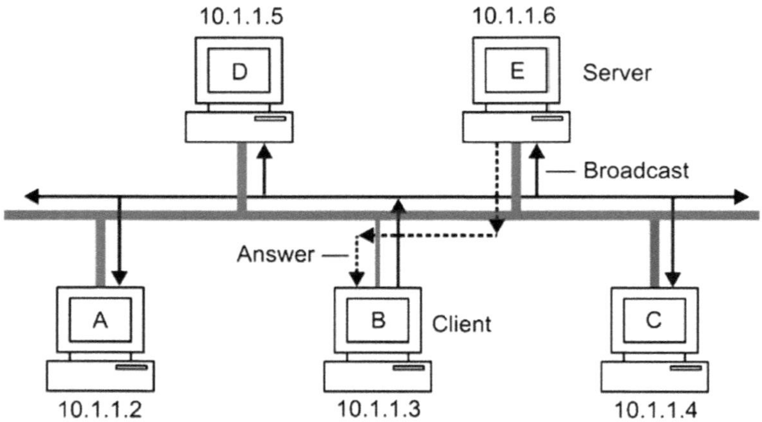

Fig. 5.49 The process of an ARP request

server's MAC address. The station with the requested IP address sends back its MAC address. From then on the mapping between the server's MAC and IP address is saved in the client's ARP table (to view the contents of the ARP table in Windows, open the *Command prompt* window and entering `arp–a`). IP packets and their known MAC addresses can now be integrated into Ethernet frames. A TCP connection is then established at the higher levels of the OSI model, after which the application layer can then send the HTTP request for the transmission of a Web site.

5.3.6.2 The Dynamic Host Configuration Protocol

IP addresses can either be manually configured or assigned automatically (DHCP). With the latter, as soon as a station (DHCP client) is switched on, it sends out a broadcast requesting an IP address. A DHCP server replies (Fig. 5.50) by assigning

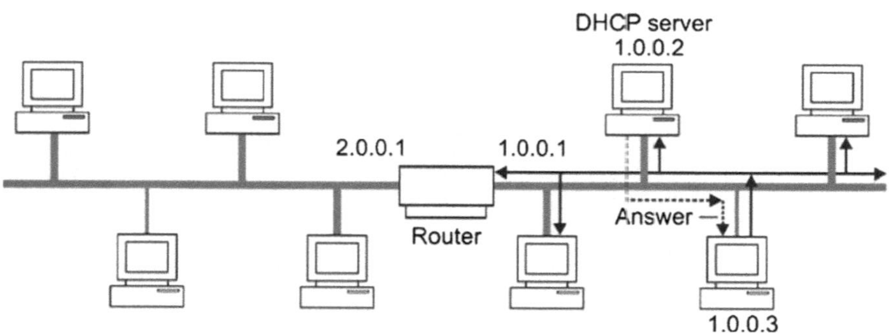

Fig. 5.50 DHCP address assignment

the station an IP address (1.0.0.3) and sending it additional information such as the network mask (255.0.0.0) and the IP address of the router (1.0.0.1) responsible for forwarding packets to other networks. The IP address is assigned either randomly from a pool of available addresses or according to a list that contains the requesting station's MAC and IP addresses.

The advantage of DHCP is that it reduces the work of the network administrator, because if changes have been made to addresses, network masks or router access, only the DHCP server needs to be reconfigured. The DHCP clients then automatically receive up-to-date information. This means that the network administrator does not need to configure the network parameters for each station on site. You can view the information sent from the DHCP server by entering `ipconfig/all` in the *Command prompt* in Windows.

DHCP is used mainly for workstation computers and mobile systems. Servers and routers, as a rule, are assigned fixed IP addresses manually. The same goes for building automation systems whose stations should naturally operate with the same configuration over a long period of time.

5.3.7 Using BACnet with Internet Protocols

Today, local networks in offices almost always use Ethernet and Internet protocols. Two or more Ethernet segments are connected via switches and IP routers. Routers connect the network(s) to the Internet, but only forward IP packets and not BACnet-over-Ethernet messages. The BACnet messages therefore need to be encapsulated in IP packets, which can then be forwarded by IP routers. There are two ways of doing this:

- Tunneling routers
- BACnet/IP.

5.3.7.1 Tunneling Routers

With a BACnet tunneling router, a BACnet/IP packet assembler-disassembler (B/IP PAD) receives the local BACnet messages, encapsulates them in IP packets using UDP as the transport protocol, and then sends them to an IP router (Fig. 5.51).

The IP router accepts these IP packets and forwards them to the B/IP PAD, where they are unpacked and sent to the local BACnet devices. The B/IP PAD has a routing table which it uses to ensure that the messages sent over the Internet end up at the right destination. This table contains BACnet network numbers, the IP addresses of the B/IP PADs in these networks and the address of the nearest IP router. The destination network number is used to determine the IP address of the appropriate B/IP PADs and a packet is then sent to it. If a global broadcast is transmitted, then all B/IP PADs entered in the routing table receive a packet.

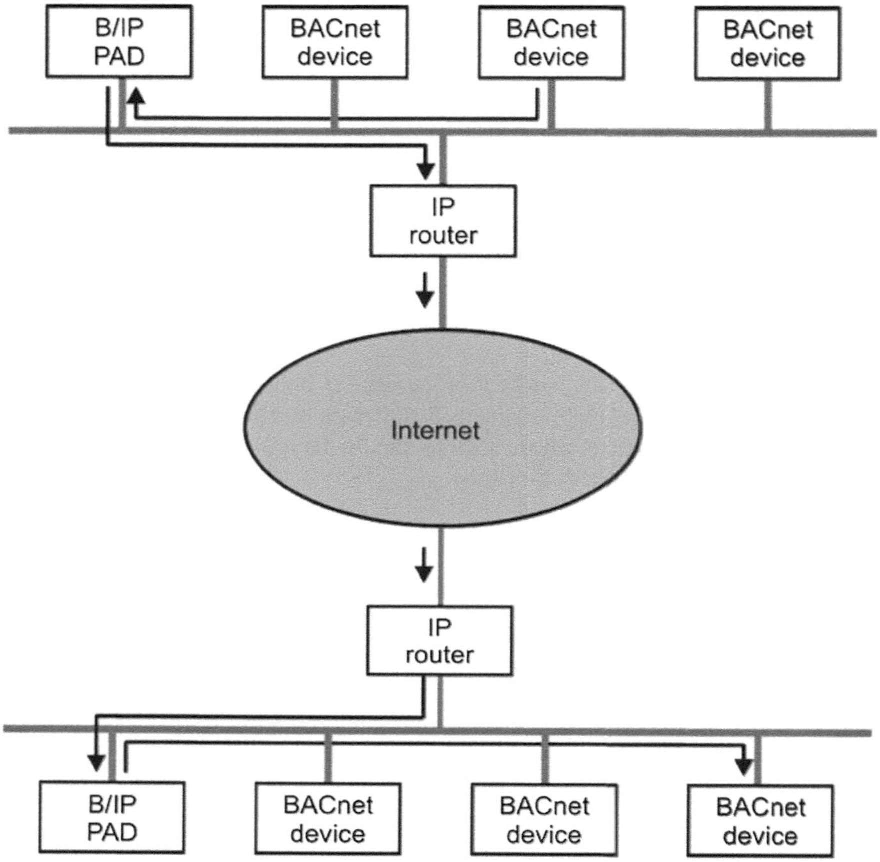

Fig. 5.51 Tunneling routers

This method is transparent for the BACnet devices in the local networks, in other words, the B/IP PAD behaves like a BACnet router that, for example, connects one Ethernet LAN with other Ethernet LANs. The BACnet devices do not see that the BACnet messages are actually transported in IP packets over the Internet.

As BACnet devices do not have to interpret IP packets, this means they tend to be relatively inexpensive. A solution with several Ethernet segments each with relatively few BACnet devices can be quite expensive, because each segment needs a B/IP PAD. Furthermore, the messages sent to other networks have to be forwarded twice within one local network (BACnet device to PAD and PAD to IP router), resulting in a high volume of traffic. As a result, the trend now is moving towards IP-compatible BACnet devices.

5.3.7.2 BACnet/IP

BACnet/IP devices are IP-compatible, which means that they can communicate with each other over the Internet (over an IP router) without B/IP PADs.

A BACnet/IP network consists of one or more IP networks or IP subnets, which are each assigned a BACnet network number. The messages are transported using IP and UDP, in which the combination of the IP address (4 bytes) and the UDP port (2 bytes) performs a similar function to a MAC address in Ethernet, MS/TP, and so on. In BACnet, IP and UDP are layer two protocols even though in the Internet they belong to OSI layers three and four. IP and UDP constitute a form of embedded protocol in BACnet. IP and UDP perform the same tasks for the BACnet network layer as the other layer two protocols, namely transporting messages between two stations and transmitting broadcasts.

That IP is a layer three protocol can be seen by its broadcasting functionality. Broadcasting to all devices in the same network is a matter of course, whereas an IP router generally does not forward a broadcast to other IP networks. Otherwise broadcasts sent to all computers worldwide can completely overload the Internet, causing it to crash. Because BACnet relies on broadcasts, for example, to identify connected devices automatically, you need in this case so-called BACnet/IP broadcast management devices (BBMDs) (see Fig. 5.52).

BBMDs listen to BACnet broadcast messages sent by BACnet devices within their networks. These messages are then sent by the BBMD in one network as targeted IP packets to the BBMDs in the other networks, which convert them into broadcasts before transmitting them. This mechanism requires that each IP network or subnet has a BBMD that has stored in its table the IP addresses of all the other BBMDs in the BACnet/IP network.

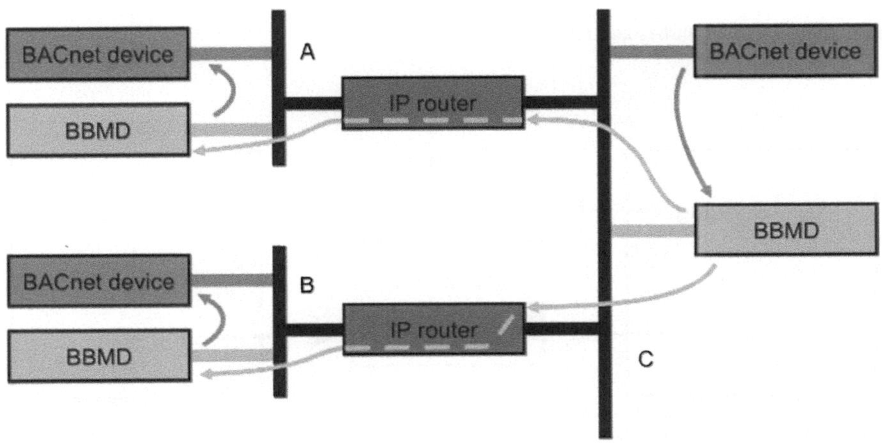

Fig. 5.52 BACnet broadcast management devices forward broadcasts to other IP networks

5.4 The Application Layer

5.4.1 Data Unit and Purpose

The application layer deals with the transfer of information to do with, for example, control, monitoring, requests and alerts. The information is transferred in the form of coded command sequences in the data section of an application protocol data unit (APDU) (Fig. 5.53).

The following sections will explain the concepts of the object types, services and procedures involved.

The BACnet application layer also performs tasks normally associated with the OSI transport layer. The corresponding information is found in the APDU's header. For example, "acknowledge" services require the receiver to respond once it has received a message (Fig. 5.54).

The requesting station (client) must receive acknowledgement of receipt from the server within a certain period of time, otherwise it will resend the message. The application program is only informed when the client has still not received confirmation after resending the message several times. For a more detailed description of APDU coding and the associated flow diagrams showing the data transfer, refer to the BACnet Standard documentation [4].

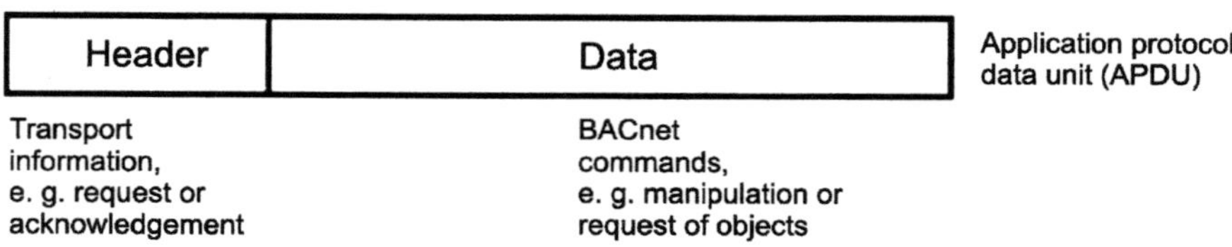

Fig. 5.53 APDU with header and data

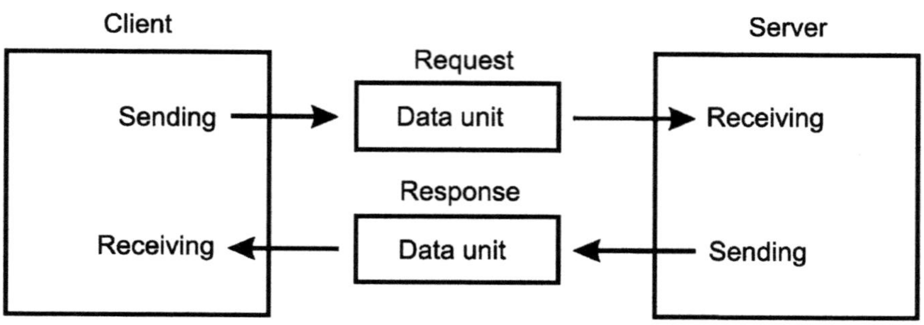

Fig. 5.54 The relationship between a client and a server

5.4.2 *Objects*

BACnet devices store information such as the temperature reading from a sensor, the switching state of a pump or a calendar showing public holidays. For the designer of a building automation system, it is not important how and where the information is stored in the device, for instance, with what operating system and in which microcontroller/memory. He just wants to know that there is a standardized way he can access this information over the Internet. To achieve this, an object-oriented approach was selected.

5.4.2.1 What is an Object?

An object is an abstract data structure in which information is stored as object properties. A simple way to imagine an object is as a table with two columns.

Table 5.7 shows an example of an object for a temperature sensor. The object name is "Room temperature" and the object type is "Analog input". The current temperature can be read from the object property "Current value". Other object properties include an upper and a lower limit value, which if exceeded can trigger an alarm.

The user cannot see how the software and hardware is used to check the status of the temperature sensor. The object itself cannot be seen by the user.

BACnet objects are the basis for the functions of building automation. The BACnet standard defines object types that cover the typical demands of building automation. As well as simple analog and digital input/output objects, there are also complex objects for controlling systems, work calendars and recording trends.

You can differentiate between properties that are required to be present and readable (R), present, readable and writable (W), and optional properties (O). You can also combine optional properties with R and W.

Developers are also free to define additional non-standard object types or properties if required. This means that innovative developments can be accommodated without waiting for the standard to be amended.

Table 5.7 Representation of a BACnet object using a table

Object name	Room temperature
Object type	Analog input
Current value	25.3
Status	Normal
Upper limit	35.0
Lower limit	0.0

5.4.2.2 Data Types

The information stored in the properties can be any of the data types shown in Table 5.8. Other data types can be based on these basic data types. A *BACnetArray* consists of, for example, an ordered sequence of elements of the same data type. Each element in this sequence can be accessed using an index. If no index value is given, the complete sequence is returned when read. The index value "0" returns the number of entries.

The *List* data type can also be used to store a sequence of elements with the same data type. However, you cannot access individual elements or request the number of elements.

For example, the *BACnetReliability* data type comprises enumerated entries but can only accept the following values:

- 0 (NO_FAULT_DETECTED)
- 1 (NO_SENSOR)
- 2 (OVER_RANGE)
- 3 (UNDER_RANGE)

For many other BACnet data types, refer to the BACnet standard.

5.4.2.3 Naming Conventions and Address Assignment

Each object in a BACnet device is referenced by a unique 32-bit numerical *Object_Identifier* property. This identifier comprises a 10-bit long object type and a 22-bit long instance number (Fig. 5.55).

Table 5.8 Data types for BACnet objects

Data type	Content
NULL	Octet with value 0
Boolean	Boolean (logical) value, TRUE or FALSE
Unsigned integer	Positive whole numbers, for example, with 8, 16, or 32 bit long
Signed integer	Signed integer in two's complement notation, for example, 8, 16, or 32 bit long
Real	IEEE-754 simple precision floating point, for example 1.234×10^7
Double	IEEE-754 double precision floating point
Octet string	String of octets
Character string	String of characters
Bit string	Sequence of bits
Enumerated	List of binary numbers
Date	Date showing the day, month, year and the day of the week
Time	24-h time format

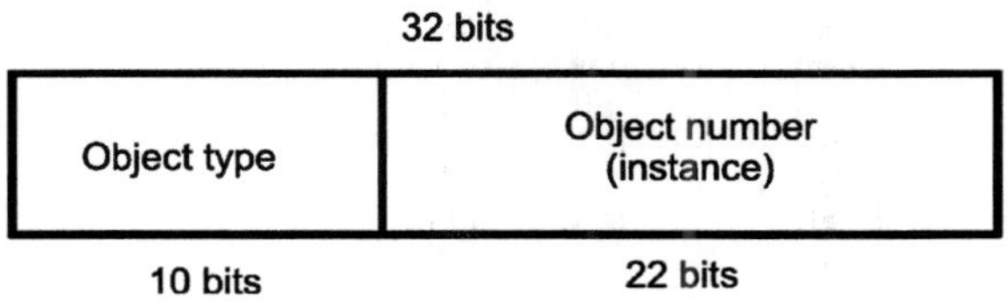

Fig. 5.55 An object identifier

When a message is sent, the object identifier needs to be entered in the data portion of the APDU in order to reference the desired object. There can be more than one object with the same object type in a device. Each BACnet device also has an object for the device itself, which also has an object identifier. This device object identifier must be unique in the whole network. Using these two identifiers, you can access any object networkwide. BACnet devices can also be implemented as a "virtual device" in the software, which means that an automation station (physical device) can contain several of these devices.

Numerical *Object_Identifier* properties are useful for automatically accessing objects. Users and programmers, however, prefer using proper names. For this reason, each BACnet object also has an object name. The device's object name must also be unique within the whole BACnet network.

A BACnet system can contain up to 2^{22} = 4,194,304 addressable devices (device object type) corresponding to the 22 remaining bits of the object identifier. At the same time, a BACnet system can contain up to 65,535 interconnected networks.

Tables 5.9 and 5.10 show a system for assigning network numbers. Buildings and floors can be determined directly from the network number.

Table 5.9 Assigning network numbers

B	B	B	F	F	
BBB:		Numbers 001–655 are assigned to each building			
FF:		00 for the building backbone network, 01–35 indicating the floors or separate systems in a building			

For example: 54,321 for building 543 and floor 21

Table 5.10 Assigning device object identifiers

X	X	F	F	B	B	B	
XX:		Numbers 00–40					
FF:		00 for the building backbone network, 01–35 indicating the floors or separate systems in the building					
BBB:		Numbers 001 to 655 are assigned to each building					

For example: 1,234,567 for device number 12 on floor 34 in building 567

This network number is also to be found in the device object identifier. The last number (BBB) is the number of the building, the middle number indicates the floor or building backbone network (FF), and the first number between 00 and 40 is for each device (XX).

This numbering system has the advantage of providing a great deal of information for someone troubleshooting communication problems. This is because the physical location (building, floor) of the device can be deduced from the device object identifier. This hierarchical addressing approach means that there can be a maximum of 41 devices on a floor, 35 floors in a building, and a total of 655 buildings. This division may not be enough for very large buildings, in which case an alternative numbering system may have to be used.

5.4.3 *Standard Object Types*

The following sections will explain the properties of the main standard object types. In the first example, the properties of the Analog Input object type are listed in a table or are explained in the description that follows. For the other object types, only their characteristic properties will be shown and explained. For more detailed information about the properties, refer to the BACnet standard.

5.4.3.1 The ANALOG_INPUT Object Type

The ANALOG_INPUT object type defines a standardized object that represents the properties of an analog hardware input. A temperature sensor could be connected to this input. If a device has more than one analog input, then you must create other instances of the same object type with different Object_Identifiers. The object properties can be requested using the BACnet services (Fig. 5.56).

Table 5.11 shows the properties for an example case. The properties are explained below.

- *Object_Identifier*: a unique numeric code (32 bit) in the device that is used to identify the object
- *Object_Name*: name of the object (the minimum length of the string shall be one character)
- *Object_Type*: type class of the object (here: Analog Input)
- *Present_Value*: current value in engineering units of the analog input being measured (The Present_Value property shall be writable when Out_Of_Service is TRUE)
- *Description*: content of the property is not restricted
- *Device_Type*: description of the sensor connected to the analog input
- *Status_Flags*: this property represents four Boolean flags that indicate the status of the analog input: {IN_ALARM, FAULT, OVERRIDDEN, OUT_OF_SERVICE}

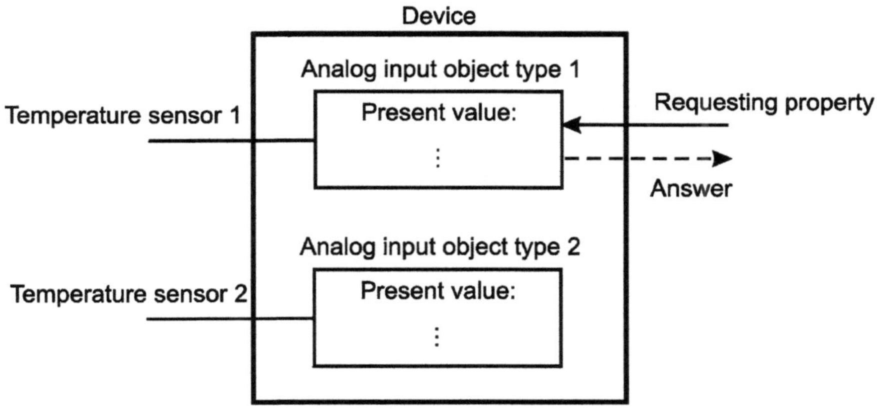

Fig. 5.56 A BACnet device with two analog input objects

- IN_ALARM: logical FALSE if the property Event_State has a value NORMAL, otherwise logical TRUE
- FAULT: logical TRUE if the Reliability property is presented and does not have a value NO_FAULT_DETECTED, otherwise logical FALSE
- OVERRIDDEN: logical TRUE if local settings or commands have been overridden. The content of the Properties Present_Value and Reliability are no longer tracking changes to the physical input, otherwise the value is logical FALSE
- OUT_OF_SERVICE: logical TRUE if the property Out_Of_Service has a value TRUE, otherwise logical FALSE

- *Event_State*: this property indicates the state of the object after an event.
- *Reliability*: this property provides an indication of whether the measurement value in the *Present_Value* property is reliable. The property may have any of the following values:

NO_FAULT_DETECTED, NO_SENSOR, OVER_RANGE, UNDER_RANGE, OPEN_LOOP, SHORTED_LOOP, UNRELIABLE_OTHER}. For example a device can have for the connected sensors an integrated cable break monitoring that in case of breakage sends an error message.

- *Out_of_Service*: this property indicates whether the input is "out of service" (TRUE) or "operational" (FALSE). If *Out_of_Service* is TRUE, the input can no longer influence the *Present_Value* property. The *Reliability* property and the corresponding FAULT *Status_Flag* are also not affected. Both values, however, may be written for testing purposes.
- *Update_Interval*: this property indicates the maximum period of time between *Present_Value* updates in hundredths of a second when the input has not been overridden or is not out of service.

Table 5.11 Properties of the ANALOG_INPUT object type

Property	Data type	R/W/O	Information
Object_Identifier	*BACnetObjectIdentifier*	R	(Analog Input, Instance 4)
Object_Name	*CharacterString*	R	"Building 007, Room 004"
Object_Type	*BACnetObjectType*	R	ANALOG_INPUT
Present_Value	*Real*	R[1]	25.3
Description	*CharacterString*	O	"Room air temperature"
Device_Type	*CharacterString*	O	"NTC Type 4711"
Status_Flags	*BACnetStatusFlags*	R	{FALSE,FALSE,FALSE,FALSE}
Event_State	*BACnetEventState*	R	NORMAL
Reliability	*BACnetReliability*	O	NO_FAULT_DETECTED
Out_Of_Service	*Boolean*	R	FALSE
Update_Interval	*Unsigned*	O	10 (seconds)
Units	*BACnetEngineeringUnits*	R	°C (measurement unit for Present_Value)
Min_Pres_Value	*Real*	O	−10.0 (lowest number of the measurement range)
Max_Pres_Value	*Real*	O	60.0 (highest number of the measurement range)
Resolution	*Real*	O	0.2 (smallest recognizable change in Present_Value in units)
COV_Increment	*Real*	O[2]	0.5 (threshold value for transmission/notification)
Time_Delay	*Unsigned*	O[3]	10 (seconds before error message is sent)
Notification_Class	*Unsigned*	O[3]	5 (Notification Class object Nr. 5)
High_Limit	*Real*	O[3]	40.0 (upper limit)
Low_Limit	*Real*	O[3]	5.0 (lower limit)
Deadband	*Real*	O[3]	2.0 (range between the high and low limits)
Limit_Enable	*BACnetLimitEnable*	O[3]	{TRUE, TRUE} (enable reporting of all high and low limits)
Event_Enable	*BACnetEventTransitionBits*	O[3]	{TRUE, FALSE, TRUE}
Acked_Transitions	*BACnetEventTransitionBits*	O[3]	{TRUE, TRUE, TRUE}
Notify_Type	*BACnetNotifyType*	O[3]	EVENT
Event_Time_Stamps	*BACnetARRAY[3] of BACnetTimeStamp*	O[3]	((12-JAN-07,10:23:24.3), (*-*-*,*:*:*.*), (26-FEB-07,23:55:22.3)
Profile_Name	*CharacterString*	O	"HBN_Temp_Sensor"

R[1]: is writable when *Out_Of_Service* is TRUE
O[2]: is required when the object supports reporting by minimum change in *Present_Value*
O[3]: is required when the object supports intrinsic reporting

- *COV_Increment*: the change of value (COV) reporting allows a COV-client to receive automatically report (notification) of some changes of the input signal. The *COV_Increment* property specifies the minimum change in *Present_Value* property that will cause this notification.
- *Time_Delay*: this property specifies the minimum period of time that the *Present_Value* must remain outside the band defined by *High_Limit* and *Low_Limit*, so that a TO_OFFNORMAL event is generated. If the *Present_Value* remains within the same band, a TO_NORMAL event is generated.
- *Notification_Class*: specifies the notification class to be used when handling and generating event notifications for this object (e.g., notifying particular recipients).
- *Deadband*: if the *Present_Value* falls below the *High_Limit* minus the *Deadband* (resp. exceeds the *Low_Limit* plus the *Deadband*), a TO_NORMAL event is generated. This range prevents a constant change between NORMAL and OFFNORMAL state when the *Present_Value* is just moving around the *High_Limit* or *Low_Limit*.
- *Limit_Enable*: this property enables the generation of events for exceeding the *High_Limit* and *Low_Limit*.
- *Event_Enable*: this property conveys three flags that enable and disable reporting of TO_OFFNORMAL, TO_FAULT and TO_NORMAL events.
- *Acked_Transitions*: this property conveys three flags that separately indicate the receipt of acknowledgements for TO_OFFNORMAL, TO_FAULT and TO_NORMAL events.
- *Notify_Type*: this property shows whether the notifications generated by the object should either *Event* or (less importantly) *Alarm* notifications.
- *Event_Time_Stamps*: this property conveys the times of the last event notifications for OFFNORMAL, TO_FAULT und TO_NORMAL events.
- *Profile_Name*: this is an optional property that shows the particular vendor-specific object profile to which the object belongs. A profile defines additional properties that go beyond those mentioned in the standard.

5.4.3.2 The ANALOG_OUTPUT Object Type

The ANALOG_OUTPUT object type (Table 5.12) is used for output analog values (voltage, current) from a BACnet devices' hardware port. The incoming present values are stored in the *Priority_Array* property in the order of decreasing priority (see Sect. 5.4.5.2). The value with the highest priority is accepted by the *Present_Value* property. *Relinquish_Default* property is the default value if there are no command priority values. The set values can also be a percentage of the maximum value.

Table 5.12 Properties of the ANALOG_OUTPUT object type

Property	Data type	R/W/O	Information
Object_Type	*BACnetObjectIdentifier*	R	ANALOG_OUTPUT
Present_Value	*Real*	W	34.6
Units	*BACnetEngineeringUnits*	R	PERCENT
Priority_Array	*BACnetPriorityArray*	R	{NULL,NULL,NULL, NULL,34.6,...}
Relinquish_Default	*Real*	R	50.0

5.4.3.3 The ANALOG_VALUE Object Type

The ANALOG_VALUE object type supplies an analog value that is not a hardware input, but rather a section of memory residing in the BACnet device. The value can be the result of a calculation. An air-conditioning system calculates the enthalpy and stores the value in an ANALOG_VALUE object (Table 5.13).

5.4.3.4 The AVERAGING Object Type

The AVERAGING object (Table 5.14) records the minimum, maximum and average value of a property during a specific period of time. The sampled value may be the value of any object within the device or an object property in another BACnet device.

The source of the data for the AVERAGING object type is defined over the *Object_Property_Reference* property. In addition to the *Average_Value* and the optional *Variance_Value*, the minimum/maximum values and the associated times can also be stored.

The AVERAGING object uses the so-called "sliding window" technique, which maintains a buffer of n samples distributed over the specified interval ΔT. The *Window_Interval* property indicates the period of time (500 s), while the *Window_Samples* property indicates the number of samples n (20) to be taken during this period of time. The *Valid_Samples* property indicates the number of samples that have been successfully collected, while the *Attempted_Samples* property indicates the number of attempted samples during the current window.

Table 5.13 Properties of the ANALOG_VALUE object type

Property	Data type	R/W/O	Information
Object_Type	*BACnetObjectIdentifier*	R	ANALOG_VALUE
Present_Value	*Real*	R	20.5
Description	*CharacterString*	O	"Calculating the enthalpy"
Units	*BACnetEngineeringUnits*	R	Joules per kilogram of dry air

Table 5.14 Properties of the averaging object type

Property	Data type	R/W/O	Information
Object_Type	*BACnetObjectIdentifier*	R	AVERAGING
Minimum_Value	*Real*	R	3.5
Minimum_Value_Timestamp	*BACnetDateTime*	O	(23-MAR-2007,13:23:45.33)
Average_Value	*Real*	R	24.3
Variance_Value	*Real*	O	8.6
Maximum_Value	*Real*	R	35.3
Maximum_Value_Timestamp	*BACnetDateTime*	O	(23-MAR-2007,14:05:22.31)
Attempted_Samples	*Unsigned*	W	20
Valid_Samples	*Unsigned*	R	19
Object_Property_Reference	*BACnetDeviceObject-PropertyReference*	R	(Analog input instance 4)
Window_Interval	*Unsigned*	W	500
Window_Samples	*Unsigned*	W	20

Normally both the *Valid_Samples* and *Attempted_Samples* properties contain the same value as *Windows_Samples* property. If *Attempted_Samples* is less than *Window_Samples*, this means that the BACnet device was restarted or device settings (e.g., *Window_Samples*) were changed. If the *Attempted_Samples* is larger than *Valid_Samples*, this means that some of the samples have gone missing.

5.4.3.5 The BINARY_INPUT Object Type

A BINARY_INPUT object type (Table 5.15) defines a standardized object whose properties represent the characteristics of a binary output. A "binary output" is a physical device or hardware input that can be in only one of two states (on/off, 1/0, active/inactive, etc.). Binary inputs are typically used for indicating whether mechanical equipment such as a pump or fan is running or is idle.

In some applications, electronic circuits may reverse the relationship between the application-level logical states ACTIVE and INACTIVE and the physical state of

Table 5.15 Properties of the BINARY_INPUT object type

Property	Data type	R/W/O	Information
Object_Type	*BACnetObjectIdentifier*	R	BINARY_INPUT
Present_Value	*BACnetBinaryPV*	R	ACTIVE
Polarity	*BACnetPolarity*	R	NORMAL
Change_Of_State_Time	*BACnetDateTime*	O	(15-JAN-2007,06:44:12.3)
Change_Of_State_Count	*Unsigned*	O	34

the underlying hardware. For example, a normally open relay contact may result in an ACTIVE state when the relay is energized, while a normally closed relay contact may result in an INACTIVE state when the relay is energized. The BINARY_INPUT object has a *Polarity* property that enables this inversion.

Other additional properties help simplify analysis and diagnosis. The *Change_Of_State_Time* property represents the date and time at which the most recent change of state occurred. *Change_of_State_Count* indicates the number of changes of state since the last time the counter was reset. The number of operating hours can also be recorded using the *Elapsed_Active_Time* property.

5.4.3.6 The BINARY_OUTPUT Object Type

The BINARY_OUTPUT object type (Table 5.16) is used to switch component such as fan that is connected to a BACnet device on or off. The two distinct states possible are ACTIVE and INACTIVE. As with a Binary Input object, the *Polarity* property can be used to invert the signal depending on the hardware used (normally open or normally closed). The *Minimum_Off_Time* and *Minimum_On_Time* properties are optional and can be used to set minimum amount of time that is to elapse before the device can be switched to the other state. This is important, particularly for motors that are not allowed to be switched on or off repeatedly in a row.

5.4.3.7 The BINARY_VALUE Object Type

Like the ANALOG_VALUE object type, the BINARY_VALUE object type (Table 5.17) is not a hardware port but an area of memory that a program can access. For example, it can be used for switching a ventilation system on or off. A program for controlling a ventilation system reads the binary value from the memory and, subject to other parameters, switches the system on or off. If a fire breaks out, the program can still let the ventilation system run, even if the *Present_Value* is set to INACTIVE.

Table 5.16 Properties of the BINARY_OUTPUT object type

Property	Data type	R/W/O	Information
Object_Type	BACnetObjectIdentifier	R	BINARY_OUTPUT
Present_Value	BACnetBinaryPV	R	INACTIVE
Polarity	BACnetPolarity	R	REVERSE
Minimum_Off_Time	Unsigned32	O	5
Minimum_On_Time	Unsigned32	O	15

Table 5.17 Properties of the BINARY_VALUE object type

Property	Data type	R/W/O	Information
Object_Type	*BACnetObjectIdentifier*	R	BINARY_VALUE
Present_Value	*BACnetBinaryPV*	R	ACTIVE

5.4.3.8 The CALENDAR Object Type

The CALENDAR object type (Table 5.18) can be used to create a list of calendar dates, in which public holidays and vacation days can be stored. The functions that are to be executed on these days are stored in a SCHEDULE object that is linked to the CALENDAR object.

5.4.3.9 The COMMAND Object Type

The COMMAND object type (Table 5.19) is used to write several properties to a group of different objects at the same time. This means you can create a command to set the whole building to one particular state. For example, a building might have two states: OCCUPIED and UNOCCUPIED. Each of these states has specific temperature and lighting parameters. A COMMAND object is normally passive. Its *In_Process* property is FALSE. When the *Present_Value* is written, the COMMAND object is active (In_Process = TRUE) and it begins processing a sequence of actions, specified in the *Action* property. The *Action* property might contain several action lists. The value of the *Present_Value* property determines which sequence of actions the COMMAND object shall take. The act of writing triggers the actions itself, which is why a list of actions can be performed in series.

However, the restart is possible only when the *In_Process* property is returned to FALSE, which means the COMMAND object is switched to a passive state. There is no guarantee that every "write" action in the action list will be successful. If any of the writes fail, then *All_Writes_Successful* is set to FALSE. According to the state of the *Quit_on_Failure-Flag* of the action lists writing can be stopped or processed to completion. But none of succeeded writes can be cancelled.

Table 5.18 Properties of the CALENDAR object type

Property	Data type	R/W/O	Information
Object_Type	*BACnetObjectIdentifier*	R	CALENDAR
Present_Value	*Boolean*	R	TRUE
Date_List	*List of BACnetCalendarEntry*	R	(((23-DEC-2006)-(6-JAN-2007)), (1-APR-2007), (1-MAY-2007))

Table 5.19 Example of properties of a COMMAND object type

Property	Data type	R/W/O	Information
Object_Type	*BACnetObjectIdentifier*	R	COMMAND
Present_Value	*Unsigned*	W	2
In_Process	*Boolean*	R	FALSE
All_Writes_Successfull	*Boolean*	R	TRUE
Action	*BACnetARRAY[N] of BACnetActionList*	R	{((,(Analog value, Instance 4), Present_Value,,23.5,,TRUE,TRUE), (,(Binary Output, Instance 2), Present_Value, ACTIVE,5,1,TRUE, TRUE)), ((,(Analog Value, Instance 4), Present_Value,,18.5,,TRUE, TRUE), (,(Binary Output, Instance 2), Present_Value, INACTIVE,5,2, TRUE,TRUE))}
Action_Text	*BACnetARRAY[N] of CharacterString*	O	{"OCCUPIED,""UNOCCUPIED"}

Table 5.20 shows the structure of an action list. The *Post_Delay* property represents a delay after the execution of each "write". If a *Device_Identifier* is not given, then command refers to an object in the same device.

The COMMAND object is a powerful tool but can cause problems if improperly configured. The *In_Process* property prevents the COMMAND object from commanding itself. With several objects, however, circular referencing could occur, in other words, a COMMAND object commands another COMMAND object that commands the first, and so on. This can cause oscillations and instability in the building automation system.

Table 5.20 Structure of an action list

Parameters	Data type
Device_Identifier (optional)	*BACnetObjectIdentifier*
Object_Identifier	*BACnetObjectIdentifier*
Property_Identifier	*BACnetPropertyIdentifier*
Property_Array_Index (if necessary)	*Unsigned*
Property_Value	*Depends on the object*
Priority (if necessary)	*Unsigned*
Post_Delay (optional)	*Unsigned*
Quit_On_Failure	*Boolean*
Write_Successful	*Boolean*

5.4.3.10 The DEVICE Object Type

Each BACnet device contains exactly one DEVICE object. The DEVICE object type (Table 5.21) defines the properties of the device and the number of available objects. The *System_Status* property reflects the current operating status of the BACnet device. Further properties define a unique vendor identification code, the model of the BACnet device, as well as the version of the firmware and application software.

5.4.3.11 The EVENT_ENROLLMENT Object Type

The EVENT_ENROLLMENT object type (Table 5.22) is used to monitor the properties of the objects and manage events (see Sect. 5.4.4.2). In addition, the property to be monitored must be defined in the *Object_Property_Reference* property. The *Event_Type* property determines which algorithm is to be used for recognizing events, for example:

- OUT_OF_RANGE
- CHANGE_OF_VALUE
- CHANGE_OF_BITSTRING

Each event type has its specific states and parameters (see Table 5.23).

Table 5.21 Example of the properties for a DEVICE object

Property	Data type	R/W/O	Information
Object_Identifier	*BACnetObjectIdentifier*	R	(Device, Instance 1)
Object_Type	*BACnetObjectType*	R	DEVICE
System_Status	*BACnetDeviceStatus*	R	OPERATIONAL
Vendor_Name	*CharacterString*	R	"HBN Inc."
Vendor_Identifier	*Unsigned16*	R	4711
Model_Name	*CharacterString*	R	"HBNBAC"
Firmware_Revision	*CharacterString*	R	"1.0"
Application_Software_Version	*CharacterString*	R	"1.0"
Protocol_Version	*Unsigned*	R	1
Object_List	*BACnetARRAY[N] of BACnetObjectIdentifier*	R	((Analog Input, Instance 1), (Analog Input, Instance 2), (Binary Input, Instance 1)
Local_Time	*Time*	O	13:22:45.34
Local_Date	*Date*	O	20-FEB-2007, TUESDAY

Table 5.22 Example of properties of the EVENT_ENROLLMENT object type

Property	Data type	R/W/O	Information
Object_Type	*BACnetObjectIdentifier*	R	EVENT_ENROLLMENT
Event_Type	*BACnetEventType*	R	CHANGE_OF_VALUE
Event_Parameters	*BACnetEventParameter*	R	(10, 0.5)
Notify_Type	*BACnetNotifyType*	R	ALARM
Object_Property_Reference	*BACnetDeviceObject PropertyReference*	R	((Device Instance 6), (Analog Input Instance 2), (Present_Value))
Event_State	*BACnetEventState*	R	NORMAL
Event_Enable	*BACnetEventTransitionBits*	R	(TRUE,TRUE,FALSE)
Notification_Class	*Unsigned*	O	3

Table 5.23 Example an event object with its state and parameters

Event type	Event state	Event parameter
CHANGE_OF_VALUE	NORMAL	Time_Delay
	OFFNORMAL	List_Of_Values

In the example in Table 5.23, the *Event_Enable* property is be used to enable notifications for the TO_OFFNORMAL and TO_NORMAL transitions. The *Event_Parameters* property enables the Time_Delay and List_Of_Values parameters to be set for the CHANGE_OF_VALUE algorithm. The notifications themselves are sent to a *Notification_ Class* object, which then forwards them to the desired destination(s). Alternatively, optional properties can be used to also send notifications to processes in BACnet devices. For more details, refer to the BACnet standard.

5.4.3.12 The FILE Object Type

The FILE object type (Table 5.24) is used to describe the properties of files that can be accessed using BACnet services.

The *File_Type* property can be used to define the type of file and its intended use. Other properties include the size of the file in bytes (*File_Size*), the last time the object was modified (*Modification_Date*) and whether or not the file data may be changed (*Read_Only* = TRUE or *Read_Only* = FALSE). The *Archive* property shows whether the file has been saved for backup purposes. The *File_Access_Method* property shows the type of file access method used, either RECORD_ACCESS for records or STREAM_ACCESS for bit-by-bit reading and writing.

Table 5.24 Example of properties for a FILE object

Property	Data type	R/W/O	Information
Object_Type	*BACnetObjectIdentifier*	R	FILE
File_Type	*CharacterString*	R	"Trend"
File_Size	*Unsigned*	R	1234
Modification_Date	*BACnetDateTime*	R	(4-APR-2006,03:35:55.3)
Archive	*Boolean*	W	FALSE
Read_Only	*Boolean*	R	FALSE
File_Access_Method	*BACnetFileAccessMethod*	R	RECORD_ACCESS

5.4.3.13 The GROUP Object Type

The GROUP object type (Table 5.25) defines a standardized object whose properties represent a collection of other objects and one or more of their properties. A GROUP object is used to simplify the exchange of information between BACnet devices.

The *List_Of_Group_Members* property lists all of the referenced objects with the chosen properties. The current values are listed in the *Present_Value* property.

5.4.3.14 The LIFE_SAFETY_POINT Object Type

The LIFE_SAFETY_POINT object (Table 5.26) is used for safety and security applications such as fire and smoke alarms.

Table 5.25 Example of a GROUP object

Property	Data type	R/W/O	Information
Object_Type	*BACnetObjectType*	R	GROUP
List_Of_Group_Members	List of ReadAccessSpecification	R	(((Analog Input, Instance 2), Present_Value, Reliability, Description), ((Analog Input, Instance 5),Present_Value, Reliability, Description)))
Present_Value	List of ReadAccessResult	R	(((Analog Input, Instance 2), Present_Value,23.5, Reliability, NO_FAULT_DETECTED, Description, "Room1"), ((Analog Input, Instance 5), Present_Value,35.4 Reliability, NO_FAULT_DETECTED, Description, "Room2")))

Table 5.26 Example of a LIFE_SAFETY_POINT object

Property	Data type	R/W/O	Information
Object_Type	*BACnetObjectIdentifier*	R	LIFE_SAFETY_POINT
Present_Value	*BACnetLifeSafetyState*	R	ALARM
Tracking_Value	*BACnetLifeSafetyState*	O	ALARM
Out_Of_Service	*Boolean*	R	FALSE
Mode	*BACnetLifeSafetyMode*	W	ON

The *Present_Value* property defines the state of the object, for example, ALARM, FAULT, BLOCKED. The exact meaning must be determined for each project. If the *Present_Value* property ends up in a "non-NORMAL" state, it remains so until it is reset. The *Tracking_Value* property, on the other hand, continuously tracks changes in the state and therefore always shows the latest state. If you set the *Out_Of_Service* property to *TRUE*, then the *Present_Value* is writable and can therefore simulate, for example, a dangerous situation. The *Mode* property conveys the desired operating mode (e.g., ON, OFF, TEST) for the LIFE_SAFETY_POINT object.

Depending on the state and parameters of other properties, events can be triggered and automatically forwarded to other BACnet devices or to a control room. For more detailed information on the comparatively complex parameter options and conditions for triggering events, refer to the BACnet standard.

5.4.3.15 The LIFE_SAFETY_ZONE Object Type

The LIFE_SAFETY_ZONE object (Table 5.27) combines several LIFE_SAFETY_POINT and LIFE_SAFETY_ZONE objects into one group for fire, safety and security applications. The *Zone_Members* property contains a list of objects that belong to the Life Safety Zone.

Table 5.27 Example of a LIFE_SAFETY_ZONE object

Property	Data type	R/W/O	Information
Object_Type	*BACnetObjectIdentifier*	R	LIFE_SAFETY_ZONE
Present_Value	*BACnetLifeSafetyState*	R	PREALARM
Tracking_Value	*BACnetLifeSafetyState*	O	PREALARM
Zone_Members	*List of BACnetDeviceObjectReference*	R	((Life Safety Point,Instance 3), (Life Safety Point,Instance 5))

5.4.3.16 The LOOP Object Type

A LOOP object is a proportional integral derivative (PID) controller whose control parameters and input/output signals can be set over BACnet. The controller requires a setpoint, which it compares with the controlled variable to calculate and display the manipulated variable using an algorithm (Fig. 5.57).

As well as the LOOP object, a control circuit also needs an ANALOG_VALUE object, an ANALOG_INPUT object and an ANALOG_OUTPUT object for the input and output values. Table 5.28 shows the most important properties.

5.4.3.17 The MULTISTATE_INPUT Object Type

This object (Table 5.29) allows the representation of discrete states. For example, binary inputs can be combined and the resulting combinations or states can be called up. How these states are created or calculated is internally defined in the BACnet device. The *Present_Value* property contains an integer number representing the state. The description belonging to this state is stored in the *State_Text* property.

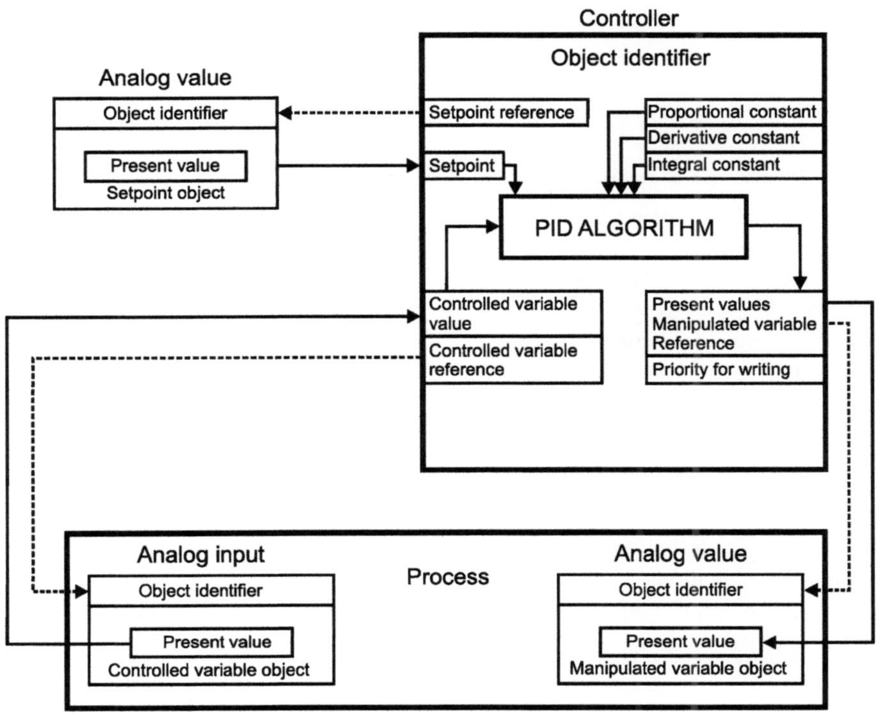

Fig. 5.57 A LOOP object and its referenced objects

Table 5.28 Example of a loop object

Property	Data type	R/W/O	Information
Object_Type	*BACnetObjectType*	R	LOOP
Manipulated_Variable_Reference	*BACnetObjectPropertyReference*	R	((Analog Output, Instance 4), Present_Value)
Controlled_Variable_Reference	*BACnetObjectPropertyReference*	R	((Analog Input, Instance 2), Present_Value)
Controlled_Variable_Value	*Real*	R	12.3
Setpoint_Reference	*BACnetSetpointReference*	R	((Analog Value, Instance 1), Present_Value)
Setpoint	*Real*	R	15.0
Proportional_Constant	Real	O	10.0
Proportional_Constant_Units	*BACnetEngineeringUnits*	O	VOLTS
Integral_Constant	*Real*	O	5.0
Integral_Constant_Units	*BACnetEngineeringUnits*	O	AMPERES
Derivative_Constant	*Real*	O	0.0
Derivative_Constant_Units	*BACnetEngineeringUnits*	O	NO_UNITS
Maximum_Output	*Real*	O	20.0
Minimum_Output	*Real*	O	2.0

Table 5.29 Example of a MULTISTATE_INPUT object

Property	Data type	R/W/O	Information
Object_Type	*BACnetObjectIdentifier*	R	MULTISTATE_INPUT
Present_Value	*Unsigned*	R	2
Number_Of_States	*Unsigned*	R	3
State_Text	*BACnetARRAY[N] of CharacterString*	O	("Hand," "Off," "Auto")

5.4.3.18 The MULTISTATE_OUTPUT Object Type

This object (Table 5.30) can be used to control one or more binary outputs. The possible combinations are represented by a state defined in the *Present_Value* property. The actual functions associated with the state are defined in the BACnet device.

Table 5.30 Example of a MULTISTATE_OUTPUT object

Property	Data type	R/W/O	Information
Object_Type	*BACnetObjectIdentifier*	R	MULTISTATE_OUTPUT
Present_Value	*Unsigned*	W	1
Number_Of_States	*Unsigned*	R	3
State_Text	*BACnetARRAY[N] of CharacterString*	O	("off," "level 1," "level 2")

5.4.3.19 The MULTISTATE_VALUE Object Type

This object (Table 5.31) can represent several binary inputs or outputs that are only assigned a section of memory in the BACnet devices as opposed to a physical port. Other BACnet devices or gateways can also access these storage locations.

5.4.3.20 The NOTIFICATION_CLASS Object Type

The NOTIFICATION_CLASS object (Table 5.32) is used for distributing event notifications to specific destinations within a BACnet system. Each of these objects is identified by a number (*Notification_Class*), which other objects can use to refer to this object. Three events: TO_OFFNORMAL, TO_FAULT and TO_NORMAL can be assigned different levels of priority using the *Priority* property. The *Recipient_List* property contains the objects that are to be notified when a particular event occurs. The notifications can also be sent to various destinations at different times. The *Ack_Required* property is used if the destination is to acknowledge receipt of the notification (Table 5.33).

5.4.3.21 The PROGRAM Object Type

The PROGRAM object can be used to control proprietary application programs in a BACnet device. As an example, we will calculate the maximum of two input values (Table 5.34).

Table 5.31 Example of a MULTISTATE_VALUE object

Property	Data type	R/W/O	Information
Object_Type	*BACnetObjectIdentifier*	R	MULTISTATE_VALUE
Present_Value	*Unsigned*	R	2
Number_Of_States	*Unsigned*	R	4
State_Text	*BACnetARRAY[N] of CharacterString*	O	("level 1," "level 2," "level 3," "level 4")

Table 5.32 Example of a NOTIFICATION_CLASS object

Property	Data type	R/W/O	Information
Object_Type	*BACnetObjectIdentifier*	R	NOTIFICATION_CLASS
Notification_Class	*Unsigned*	R	5
Priority	*BACnetARRAY[3] of Unsigned*	R	(2,4,5)
Ack_Required	*BACnetEventTransitionBits*	R	(TRUE,TRUE,TRUE)
Recipient_List	*List of BACnetDestinations*	R	(((Monday, Tuesday, Wednesday), 5:00, 19:00, (Device, Instance 23), 3, TRUE, (TRUE, TRUE, FALSE)), ((Monday, Tuesday, Wednesday),19:00, 5:00, (Device, Instance 28),3,TRUE, (TRUE, TRUE, FALSE)),

Table 5.33 Format of the *Recipient*_List property

Parameter	Data type	Description
Valid Days	*BACnetDaysOfWeek*	Days of the week on which the notifications are to be distributed to the recipients
From Time, To Time	*Time*	The window of time during which the destination is viable on the days of the week Valid Days
Recipient	*BACnetRecipient*	The destination device(s) to receive notifications
Process Identifier	*Unsigned32*	The ID of a process within the recipient device that is to receive the event notification
Issue Confirmed Notifications	*Boolean*	TRUE for confirmed notifications, FALSE for unconfirmed notifications
Transitions	*BACnetEvent TransitionBits*	A set of three flags that indicate those transitions {OFFNORMAL, TO_FAULT or TO_NORMAL} for which this recipient is suitable

Table 5.34 Example of a PROGRAM object

Property	Data type	R/W/O	Information
Object_Type	*BACnetObjectIdentifier*	R	PROGRAM
Program_State	*BACnetProgramState*	R	RUNNING
Program_Change	*BACnetProgramRequest*	W	READY
Reason_for_Halt	*BACnetProgramError*	O	NORMAL
Instance_Of	*CharacterString*	O	"Max2Refs"
Ref1	*Real*	Proprietary	((Analog Input, Instance 1), Present_Value)
Ref2	*Real*	Proprietary	((Analog Input, Instance 2), Present_Value

The origin of the input values is defined using the proprietary properties *Ref1* and *Ref2*. The *Instance_Of* property indicates the name of the application program in the BACnet device. The *Program_Change* property is used to request changes to the operating state of the process, for example, if READY is shown, this means ready for change. The program can then be loaded (LOAD), executed if not already running (RUN), stopped (HALT), restarted at its initialization point (RESTART) or stopped and unloaded (UNLOAD). The *Program_State* property indicates the logical state of the process executing the application program, for example, RUNNING or HALTED. In the latter case, the *Reason_for_Halt* property can be used to establish why program has stopped.

5.4.3.22 The SCHEDULE Object Type

The SCHEDULE object can be used to schedule when certain actions are to take place. Table 5.35 shows an example of a BINARY_OUTPUT object, which could be used for switching a heating system on or off. The *Effective_Period* property indicates the period in which this schedule is to be executed. The *Weekly_Schedule* property can be used to set the times the system should be switched on and off on certain days. Exceptions such as public holidays are dealt with using the *Exception_Schedule* property. The *Priority_For_Writing* property is used for indicating the priority level assigned to a command.

Table 5.35 Example of a SCHEDULE object

Property	Data type	R/W/O	Information
Object_Type	*BACnetObjectIdentifier*	R	SCHEDULE
Present_Value	*Depends on the data type of the referenced object's Present_Value*	R	ACTIVE
Effective_Period	*BACnetDateRange*	R	((1-SEP-2007)-(23-DEC-2007))
Weekly_Schedule	*BACnetARRAY[7] of BACnetDailySchedule*	O	{(((8:00, ACTIVE), (18:00, INACTIVE)), and so on, for each day of the week
Exception_Schedule	*BACnetARRAY[N] of BACnetSpecialEvent*	O	{((11-NOV-2007), (12:00, INACTIVE))}
List_Of_Object _Property_References	*List of BACnetDeviceObject PropertyReference*	R	((Device, Instance 3), (Binary Output, Instance 2), *Present_Value*)
Priority_For_Writing	*Unsigned(1..16)*	R	14

5.4.3.23 The TREND_LOG Object Type

A TREND_LOG object (Table 5.36) monitors a property of a referenced object and, when predefined conditions are met, saves ("logs") the value of the property and the time in an internal buffer for subsequent retrieval. It can be used, for example, to record the temperature in a room measured by a sensor.

The *Log_DeviceObject* property specifies the source of the data. The *Log_Enable* is used to enable or disable the logging function. In addition, the *Start_Time* and *Stop_Time* properties can be used to define when logging starts and stops. The data may be logged periodically (polling) or upon a change of value (COV *Change of Value*). In the first case, the *Log_Interval* property specifies the periodic interval in hundredths of a second for which the referenced property is to be logged.

The TREND_LOG object provides memory (buffer) for storing data. The *Buffer_Size* property specifies the maximum number of records the buffer may hold. When the buffer is full, the trend logging can either be stopped (*Stop_When_Full* = *TRUE*) or can overwrite the oldest value and continue logging. The *Log_Buffer* property contains for each entry a time stamp and the four *StatusFlags* for the associated objects. The *Record_Count* property represents the number of entries in the *Log_Buffer*. A value 0 written to this property deletes all the records in the buffer.

Table 5.36 Example of a TREND_LOG object

Property	Data type	R/W/O	Information
Object_Type	*BACnetObjectIdentifier*	R	TREND_LOG
Log_Enable	*Boolean*	W	TRUE
Start_Time	*BACnetDateTime*	O	(4-JAN-2007,03:35:55.3)
Stop_Time	*BACnetDateTime*	O	(14-FEB-2007,07:44:12.5)
Log_DeviceObjectProperty	*BACnetDeviceObject PropertyReference*	O	((Device Instance 7),Analog Input, Instance 3, Present_Value)
Log_Interval	*Unsigned*	O	100
Stop_When_Full	*Boolean*	R	FALSE
Buffer_Size	*Unsigned32*	R	200
Log_Buffer	*List of BACnetLogRecord*	R	((4-JAN-2007,03:35:55.3),23.5, (FALSE,FALSE,FALSE, FALSE))
Record_Count	*Unsigned32*	W	1

5.4.3.24 The TREND_LOG_MULTIPLE Object Type

This object is essentially the same as the TREND_LOG object but can monitor more properties. Using the *Log_DeviceObjectPropertyList* property instead of the Log_*DeviceObjectProperty* property means you can list all data sources. The values with their corresponding timestamps are stored in the *Log_Buffer*.

5.4.3.25 The ACCUMULATOR Object Type

The ACCUMULATOR object (Table 5.37) is used for devices that indicate measurements by counting pulses, for example, electricity or heating meters, and which are connected to a BACnet device. Each pulse represents a reduction of a specific unit of quantity. The ACCUMULATOR object measures the quantity used by counting the number of pulses.

The *Scale* property indicates the conversion factor that is multiplied with the value of the *Present_Value* property to provide a value in the units indicated by *Units*. The *Prescale* property presents the coefficients that are used for converting the pulse signals generated by the measuring instrument into the value displayed by the *Present_Value* property.

Other uses for the ACCUMULATOR type include being able to request the number of input pulses (*Pulse_Rate*) received during the most recent period (*Limit_Monitoring_Interval*). You can also specify a limit (*High_Limit*) that the *Pulse-Rate* must exceed before an event is generated and you can also log meter readings with their own time stamp.

5.4.4 BACnet Services

For communication between devices, BACnet defines services that can be used to access objects. A simple example involves requesting a value from a temperature sensor using the ANALOG_INPUT object (Fig. 5.58).

The client uses the ReadProperty service to request an object's *Present_Value* using the object's *Object_Identifier* as an address. The messages are also numbered,

Table 5.37 Example of an ACCUMULATOR object

Property	Data type	R/W/O	Information
Object_Type	*BACnetObjectIdentifier*	R	ACCUMULATOR
Present_Value	*Unsigned*	R	123
Scale	*BACnetScale*	R	2
Units	*BACnetEngineeringUnits*	R	KILOWATT_HOURS
Prescale	*BACnetPrescale*	O	(1:10)

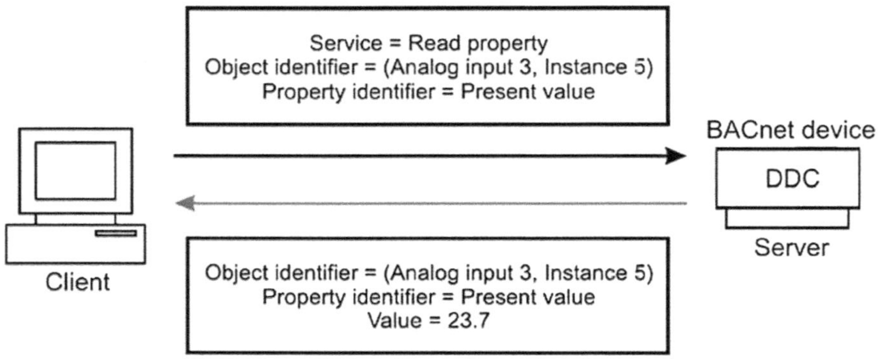

Fig. 5.58 Requesting a value from a BACnet device

which ensures that the responses can be matched to the requests. This is particularly important when there are a large number of requests. The ReadProperty service belongs to the acknowledgement services. The BACnet services are divided into the following five groups.

5.4.4.1 Object Access Services

Object access services provide the means to access and manipulate the properties of BACnet objects (Table 5.38).

Moreover, BACnet can even create new objects while in operation using the *CreateObject* service. In practice, this is used to create AVERAGING, CALENDAR, EVENT_ENROLLMENT, GROUP, NOTIFICATION_CLASS, SCHEDULE and TREND_LOG objects.

Table 5.38 Object access services

Service	Description
AddListElement	Add one or more list elements to an object property
RemoveListElement	Remove one or more elements from the property
CreateObject	Create a new instance of an object
DeleteObject	Delete an existing object
ReadProperty	Read access to any property
WriteProperty	Write access to any property of any object
ReadPropertyMultiple	Read access to any property of any object
WritePropertyMultiple	Write access to any property of any object
ReadPropertyConditional	Read access to any property of any object that meets a list of selection criteria or conditions
ReadRange	Read a specific range of data items from a list, for example, of a specific period of a TREND_LOG object

The *ReadPropertyConditional is a very powerful service*, because the request criteria can include relational operators (=, <, >), logical operators (OR, AND) and a time stamp. Using the selection criteria (*Object_Type*, = , ANALOG_INPUT), (*Present_Value*, > ,30), for example, returns all ANALOG_INPUT objects and their object identifiers whose value is greater than 30.

You can use (*Reliability*, NO_FAULT_DETECTED), (*Out_Of_Service*, = , TRUE) to locate objects that are not ready for use.

5.4.4.2 Alarm and Event Services

Alarm and Event Services can be used to communicate alerts, operating states, error messages and even simple measurements. Table 5.39 lists the services available.

The following sections will explain the three types of services available for handling events specially developed for use in building automation.

Change-of-Value Reporting

One of BACnet's strengths is its ability to automatically report changes of value, known as change of value (COV) reporting. This means that a specific standard object type (e.g., analog input/output) can be configured to send a notification when

Table 5.39 Alarm and event services

Service	Description
AcknowledgeAlarm	To acknowledge that a human operator has seen and responded to an event notification
ConfirmedCOVNotification	Confirmed notification of changes that may have occurred to the properties of a particular object
UnconfirmedCOVNotification	Unconfirmed notification of changes that may have occurred to the properties of a particular object
ConfirmedEventNotification	Confirmed notification of an event
UnconfirmedEventNotification	Unconfirmed notification of an event
GetAlarmSummary	Summary of active alarms
GetEnrollmentSummary	A summary of event-initiating objects. Different filters may be applied to define the search criteria (e.g., only objects with certain event priority)
GetEventInformation	Summary of all active event states from a device
LifeSafetyOperation	A mechanism for conveying specific instruction from a human operator to accomplish any of the following objectives: sound off/sound on, audible or visual notification appliances, reset latched notification appliances
SubscribeCOV	Subscribe for receipt of notifications of changes
SubscribeCOVProperty	Subscribe for the receipt of notifications of changes that may occur to the properties of a particular object

its present value changes by a *COV_Increment*. The same can be applied to changes of value for digital states (e.g., binary input/output). This is known as change of state (COS). Change of value shown by status flags can also trigger COV notification. The event-oriented data transmission does not overload the network because, unlike polling, the messages are only generated when needed.

A COV client (e.g., a control center) must subscribe to the COV server (BACnet device) to benefit from automatically generated reports (notifications). COV subscriptions are generated using the *SubscribeCOV* service or the *SubscribeCOVProperty* service. The subscription establishes a connection between the change-of-value detection and reporting mechanism on the COV server and a "process" in the COV client. This subscription can be permanent or temporary. However, as you cannot be sure whether a subscription is retained in the BACnet device after a power failure or restart, the subscription should be renewed regularly.

As an example of COV we will use the ANALOG_INPUT object (Table 5.11). In this example the *COV_Increment* property is set so that a notification is generated when the *Present_Value* changes by 0.5 °C. This notification is only sent to those devices that have subscribed using the *SubscribeCOV* service. The COV server saves a list of subscribers in the *Active_COV_Subscriptions* property in its device object.

Intrinsic Reporting

Intrinsic reporting is based on properties within a device that can be used to monitor alarms or events (see Table 5.40). Event notification must always be enabled (*Event_Enable*).

Refer to the temperature limit values (*High_Limit* = 40 °C, *Low_Limit* = 5 °C) set for the ANALOG_INPUT object in Table 5.11. If these limits are exceeded this leads to TO_OFFNORMAL event, and when the values return to normal this leads

Table 5.40 Example of intrinsic reporting

Object type	Criteria	Event type
Binary_Input	If *Present_Value* changes to a new state for longer than *Time_Delay*	CHANGE_OF_STATE
Analog_ Input	If *Present_Value* exceeds the range between *High_Limit* and *Low_Limit* for longer than *Time_Delay* AND the new transition is enabled in *Event_Enable* and *Limit_Enable*, OR *Present_Value* returns within the *High_Limit* – Deadband to *Low_Limit* + Deadband range for longer than *Time_Delay* AND the new transition is enabled in *Event_Enable* and *Limit_Enable*.	OUT_OF_RANGE
Loop	If the absolute difference between Setpoint and *Controlled_Variable_Value* exceeds *Error_Limit* for longer than *Time_Delay*	FLOATING_LIMIT

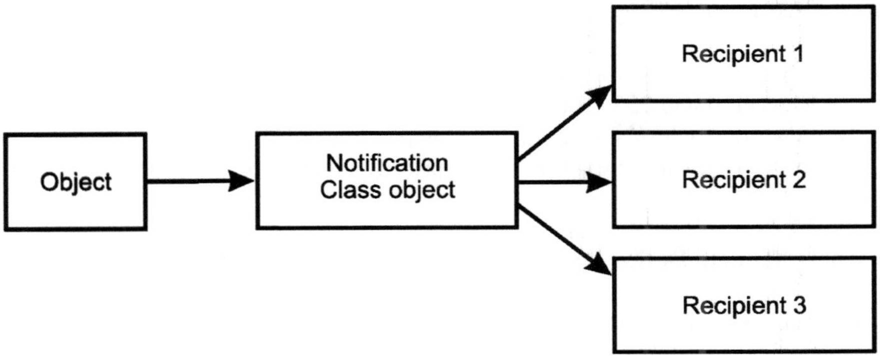

Fig. 5.59 A NOTIFICATION_CLASS object can distribute notifications to one or more recipients

to a TO_NORMAL event. Both of these events generate a notification because the *Event_Enable* property has been activated. TO_FAULT events, however, do not generate a notification. Intrinsic reporting requires another object in which to list all the notified objects (Fig. 5.59).

The NOTIFICATION_CLASS object is defined in the *Notification_Class* property of the event-initiating object. The *Notification-Class* property then lists all devices that are to be notified of the event.

Table 5.11 defines NOTIFICATION_CLASS object number 5, which in turn is described in Table 5.32.

It is often necessary to forward notifications to different destinations depending on the time and the day of the week. For example, you may want alarm notifications to be sent to control center personnel during the day. At night, however, the notifications could be sent by text message to the caretaker or to the off-duty personnel. The NOTIFICATION_CLASS object specifies the time and on which day(s) a recipient is to receive notification.

Algorithmic Change Reporting

With Algorithmic Change Reporting, a BACnet device using EVENT_ ENROLLMENT objects to determine which of an object's properties are to be monitored based on certain criteria and how notifications are to be directed to one or more destinations. As an example, refer to the EVENT_ENROLLMENT object in Table 5.22. The various algorithms for triggering notifications are described in the BACnet standard.

Alarm and Event Priority

BACnet can assign each automatically generated notification a numerical priority value between 0 and 255. The highest priority is 0 and the lowest is 255. Usually only a few priority levels are used, for example, low (255), medium (128) and high (0). Table 5.41 shows few more precise priority levels, which can also be useful.

5.4.4.3 Remote Device Management Services

These services can be used to carry out administrative tasks such as:

- Starting or stopping a BACnet device (*DeviceCommunicationControl*)
- Instructing a BACnet device to reboot or restart itself (*ReinitializeDevice*)
- Synchronizing (*TimeSynchronization*) the clock in the BACnet devices with a main clock (*Time Master*)

Furthermore, you can check what BACnet devices are present on the network or request the mapping between the *Object_Identifier* and *Object_Name*.

The *Who-Has* service (broadcast) is used to identify the BACnet devices that have a specific object identifier or object name (Fig. 5.60). The *I-Have* service is used to respond to a *Who-Has* request by supplying the device object identifier and the network address (in the network header).

The *Who-Is* service (Fig. 5.61) is used to request the network address and device object identifier from all BACnet devices. The devices respond using the *I-Am* service. If the device object identifier of a particular device is already known, then only a network address can be requested.

Table 5.41 Example of alarm and event priority levels

Notification level	Priority range	Example
Risk to life	00–31	Fire alarm
Risk to property	32–63	Security alarm (burglary, forced entry)
Supervisory	64–95	Technical alarm (heating system failure), immediate action necessary
Trouble	96–127	Alert (temperature sensor failure), no immediate action necessary
Service and maintenance	128–191	System maintenance required
Operating mode	192–255	Changing mode (day/night)

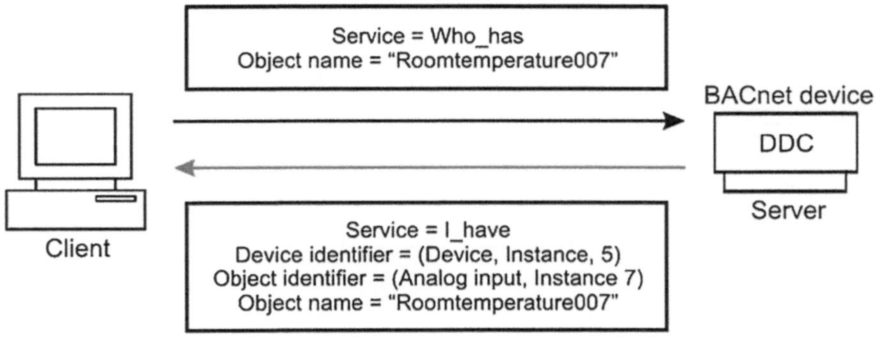

Fig. 5.60 A *Who-Has* request and an *I-Have* response

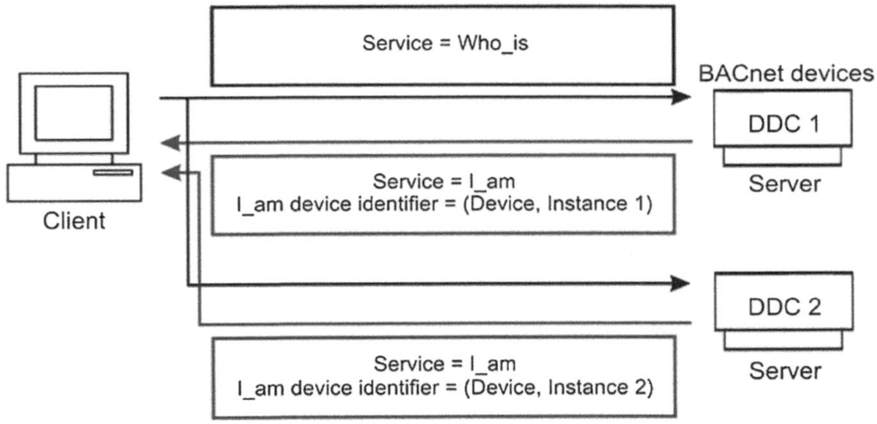

Fig. 5.61 A *Who-Is* request and an *I-Am* response

5.4.4.4 File Access Services

Files in BACnet are a collection of octets of arbitrary length and meaning. They are used particularly for vendor-specific applications. A BACnet device has a FILE object for each file that it accesses using BACnet services. The *AtomicReadFile* service is used to read from and the *AtomicWriteFile* is to write to the FILE object. Atomic in this context means that only one file access operation is allowed for the same file.

5.4.4.5 Virtual Terminal Services

The term "terminal" stems back to the time of the mainframe computer. Back then you did not have a workstation computer, just a keyboard and a monitor that were

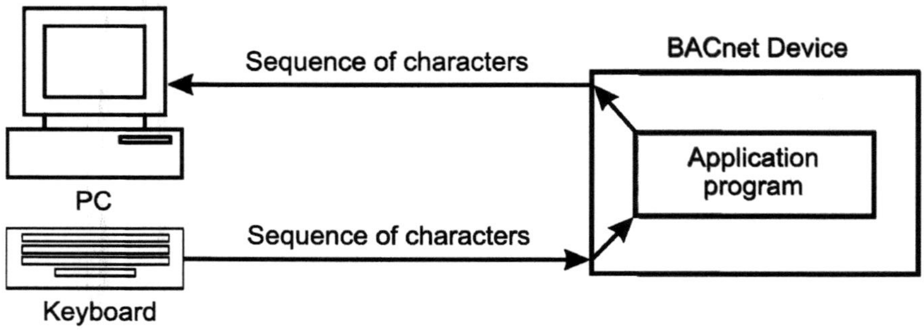

Fig. 5.62 Virtual terminal service for controlling an application program

linked to a main frame computer over a serial connection. A user's keystrokes on the keyboard were transmitted directly to the mainframe, which then sent them back to your monitor. A similar procedure to this is the TELNET protocol used on the Internet, in which commands can be sent using a text-based program to a remote computer. BACnet provides a similar service that administrators can use, for example, to configure BACnet devices (Fig. 5.62).

5.4.5 BACnet Procedures

5.4.5.1 Backup and Restore

Many BACnet devices contain configuration data that is set up by a vendor's proprietary configuration tool. This information can be lost if there is a failure or a technical malfunction, which means there must be mechanisms available for backing up and restoring the configuration data and programs. BACnet provides standardized backup and restore procedures based on the known BACnet services such as *AtomicReadFile* and *AtomicWriteFile*.

5.4.5.2 Command Prioritization

Objects in a BACnet control system can be manipulated by a number of entities. For example, a heating system could be switched on or off using a local operating panel or the main control center. A form of arbitration is therefore essential to establish who is given priority access, in other words, which has the highest priority.

BACnet has a prioritization scheme that assigns varying levels of priority. The priority level is assigned using a *Write* request to the object's properties. Only commandable properties can be assigned a priority. For ANALOG_INPUT and

Table 5.42 BACnet standard command priorities

Priority level	Application
1	Manual-Life Safety
2	Automatic-Life Safety
3	Available
4	Available
5	Critical Equipment Control
6	Minimum On/Off
7	Available
8	Manual Operator
9–16	Available

BINARY_OUTPUT object types, this is the *Present_Value* property. The *Write* request includes a conditional "Priority" parameter that ranges from 1 to 16 (Table 5.42).

How the priorities are assigned depends on the site; however, a few have been standardized (see Table 5.43). Each commandable property of an object has an associated priority table represented by the *Priority_Array* property. The *Priority_Array* consists of an array of commanded values in order of decreasing priority. The first value in the array corresponds to priority 1 (highest), the second value corresponds to priority 2, and so on, down to value 16 (the lowest priority). An entry in the *Priority_Array* property may have a commanded value or a NULL. A NULL value indicates that there is no existing command with that priority. An object continuously monitors all entries within the priority table in order to locate the entry with the highest priority non-NULL value and sets the commandable property to this value (Table 5.43).

The switch command from the operating panel (priority 8) is entered in the table, but is not executed until the entry from the control center (priority 5) has been removed (relinquished). The process of relinquishing a command is carried out by a write operation similar to the command itself, except that the commandable property value is NULL. Relinquishing a command places a NULL value in the *Priority_Array* property corresponding to the appropriate priority.

Table 5.43 Example of command prioritization

Priority level	No command	Command priority 5	Command priority 8	Command priority 8
…	Null	Null	Null	Null
5	Null	Off	Off	Null
…	Null	Null	Null	Null
8	Null	Null	On	On
…	Null	Null	Null	Null
16	Null	Null	Null	Null
Default	Off	Off	Off	Off
Present_Value	Off	Off	Off	On

Table 5.44 Example of command prioritization with minimum on/off time

Priority level	No command	Command priority 8	Command priority 7	No command
...	Null	Null	Null	Null
6	Null	On	On	Off
7	Null	Null	Off	Off
8	Null	On	ON	On
...	Null	Null	Null	Null
16	Null	Null	Null	Null
Default	Off	Off	Off	Off
Minimum_On_Time	–	Running	Running	Expired
Minimum_Off_Time	–	–	–	Running
Present_Value	Off	On	On	Off

A special feature of this priority mechanism is that it enables you to determine the minimum on and off times (Table 5.44). Some electronic motors must, for example, come to a complete halt before the rotational direction can be reversed. Changing the direction too quickly can damage the motor. As long as the minimum on/off time has not elapsed, the present value remains in position six in the priority table. A command to change direction is entered in the table but has no effect. Once the minimum on/off time has lapsed, the priority six entry is removed and the change direction command is executed. Commands with priorities higher than six are executed immediately and should be reserved for emergencies.

5.5 BACnet Devices and Interoperability

BACnet was developed as an open standardized protocol for building automation systems, the ultimate goal of which was and is the "interoperability" of devices. Interoperability is the ability of devices from different vendors to work together through the digital exchange of information. You do not have to use devices from different vendors in the same project; however, this does allow the project designer the luxury of having a wider range of products to choose from. At the same time, interoperability does not just refer to interconnecting equipment from multiple manufacturers; it also refers to the use of devices from the same manufacturer from different generations. BACnet therefore ensures security of investment and expandability of an installation, because the latest devices can be used with older ones. To meet these high demands, you need strict specifications and transparent device documentation from the vendor for the project designer.

5.5.1 *Interoperability Areas and Building Blocks*

The BACnet standard defines a number of functions for building automation that are divided into five interoperability areas (IAs). For example, there is an IA for alarm and event management and an IA for scheduling. Each IA is also assigned a number of BACnet Interoperability Building Blocks (BIBBs), which are a collection of one or more BACnet services. Figure 5.63 shows the identifiers used.

If you want to use a BACnet device to request, for example, a temperature value from a temperature sensor, the following conditions must be fulfilled: the requesting device (client) must have implemented the DS-RP-A service or BIBB (see Fig. 5.63) and the responding device must have implemented the DS-RP-B BIBB. Each communication between two devices involves a BIBB pair that defines the client and server's functionality (Fig. 5.64).

The following sections will introduce you to the interoperability areas and their range of functions.

5.5.1.1 Data Sharing

Data sharing is the exchange of information between BACnet devices.
This can be used for the following purposes:

- Displaying sensor and calculated values
- Modifying properties, setpoint and operational parameters
- Archiving operational data

Read and write services as well as event-oriented data transmission (COV) are required for data sharing. The typical BIBBs that belong to data sharing are Data Sharing ReadProperty-A (DS-RP-A) on the client side and Data Sharing ReadProperty-B (DS-RP-B) on the server side.

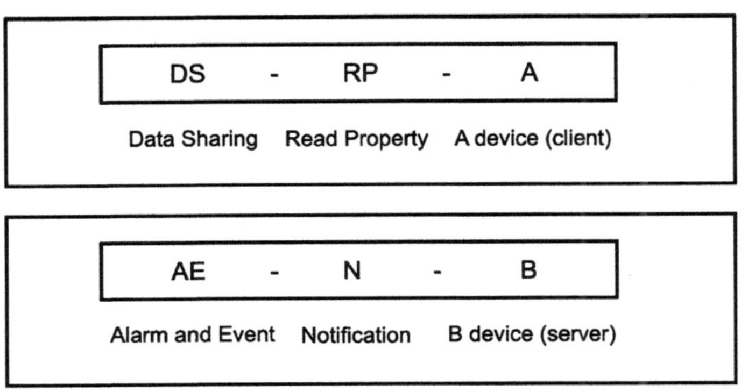

Fig. 5.63 Example of the identifiers for BIBBs

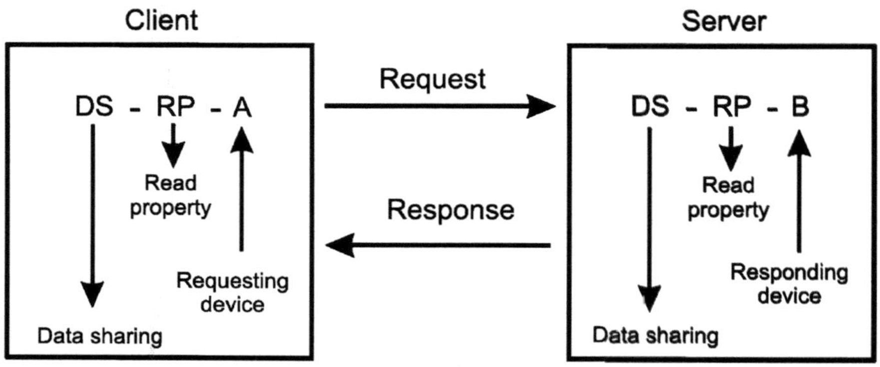

Fig. 5.64 Data sharing between BIBB pairs

5.5.1.2 Alarm and Event Management

Alarm and event management includes the services necessary to carry out functions such as:

- Notification and modification of event and alarm information
- Alarm acknowledgement
- Alarm summarization

Alarm and Event Notification A (AE-N-A) is one of the BIBBs used and enables a BACnet device to process alarm and event notifications.

5.5.1.3 Scheduling

Scheduling is the exchange of data between BACnet devices related to the establishment and maintenance of dates on and times at which specified output actions are to be taken. In order to do this, you need time schedules for storing public holidays and vacation days. This IA includes Scheduling Internal B BIBB, which provides the data and time scheduling for circuits. A BACnet device with this functionality must have at least one CALENDAR and one SCHEDULE object. The SCHEDULE object must support at least six entries a day. Naturally, DS-RP-B and DS-WP-B must also be present so that the entries in the objects can be read and written. In addition, the Device Management Time Synchronization (DM-TS-B) and Device Management UTC Time Synchronization (DM-UTC-B) BIBBS should also be supported so that the data and time in the BACnet device can be set.

5.5.1.4 Trending

Trending is the accumulation of pairs (time, values) at specific rates for a specified duration. The values can either be recorded at regular intervals or based on events. Trending Viewing and Modifying Trends Internal B (T-VMT-I-B) is an example of a trending BIBB. BACnet devices that support this BIBB can save data in an internal buffer and have at least one TREND_LOG object.

5.5.1.5 Device and Network Management

The device and network management IA defines various functions for monitoring, setting up and configuring BACnet devices, including:

- Displaying the status of any device on the BACnet internetwork
- Ability to synchronize the time in those devices that maintain clocking devices
- Restarting or rebooting BACnet devices
- Backing up and restoring data
- Establishing point-to-point connections between half-routers

For example, a BACnet device can authorize a BACnet device containing the Device Management Reinitialize Device B (DM-RD-B) BIBB to reinitialize.

5.5.2 BACnet Device Profiles

You can compare the functions of BACnet devices by comparing the BIBBs they support. Owing to the large number of BIBBs, this is not that straightforward. To simplify proceedings the BACnet devices can be divided into six groups, or BACnet device profiles. Each profile specifies the minimum requirements and capabilities to ensure interoperability. Device manufacturers are also free to implement additional services.

5.5.2.1 BACnet Operator Workstations

The BACnet Operator Workstation (B-OWS) is the user interface for the BACnet system. While it is primarily used for the operation of a system, the B-OWS may also be used for configuration activities that are beyond the scope of the BACnet standard. The B-OWS is not intended for the direct digital control of systems. But it does enable the specification of the following:

- Data Sharing (DS)
  - Archival storage of data
  - Presentation of data (i.e., reports and graphics)
  - Displaying measuring values and states
  - Setpoints and parameters modification
- Alarm and Event Management (AE)
  - Displaying alarms and events
  - Alarm acknowledgment by operators
  - Alarm summarization
  - Adjustment of alarm limits and routing
- Scheduling (SHED)
  - Modification of schedules
  - Displaying the start and stop times (schedule) of scheduled devices
- Trending (T)
  - Modification of trend log parameters
  - Displaying and archiving trend data
- Device and Network Management (DM)
  - Displaying information about the status of any device on the BACnet internetwork
  - Displaying information about any object in the BACnet internetwork
  - Ability to silence network devices that are transmitting erroneous data
  - Ability to synchronize the time in device across the BACnet internetwork
  - Ability to cause a remote device to reinitialize itself
  - Ability to backup and restore the configuration of other devices
  - Ability to command half-routers to establish, terminate and configure connections

A B-OWS must support the BIBBs listed in Table 5.45. The BACnet standard contains a list of BIBBs for each profile, including a description.

Table 5.45 BIBBs for the BACnet operator workstation sorted by IAs

IA	DS	AE	SCHED	T	DM
BIBBs	DS-RP-A,B	AE-N-A	SCHED-A	T-VMT-A	DM-DDB-A,B
	DS-RPM-A	AE-ACK-A		T-ATR-A	DM-DOB-A,B
	DS-WP-A	AR-ASUM-A		T-VMMV-A	DM-DCC-A
	DS-WPM-A	AE-ESUM-A		T-AMVR-A	DM-TS-A or DM-UTC-A
					DM-RD-A
					DM-BR-A
					NM-CE-A

A BACnet operator workstation's graphical interface, configuration and programming tools, and additional tools for integrating disparate networks vary depending on the manufacturer. Unfortunately, there is no standardized development environment available such as ETS for KNX or LonMaker for LonWorks. As a result, the practical use of BACnet is strongly influenced by proprietary tools and, to a certain extent, limits the operation of components from different manufacturers.

5.5.2.2 BACnet Building Controllers

A BACnet building controller (B-BC) is programmable automation device capable of carrying out a variety of building automation and control tasks, including:

- Data Sharing (DS)
  - Ability to provide the values of any of its BACnet objects and their properties
  - Ability to retrieve the values of BACnet objects from other devices
  - Ability to modify/write properties
- Alarm and Event Management (AE)
  - Generation of alarm/event notifications and the ability to direct them to recipients
  - Maintain a list of unacknowledged alarms/events
  - Notifying other recipients that the acknowledgment has been received
  - Adjustment of alarm/event parameters
- Scheduling (SHED)
  - Ability to schedule output actions, both in the local device and in other devices, based on date and time
- Trending (T)
  - Collection and delivery of (time, value) pairs
- Device and Network Management (DM)
  - Ability to respond to status queries
  - Ability to respond to requests for information about any of its objects
  - Ability to respond to communication control messages
  - Ability to synchronize its internal clock upon request
  - Ability to perform re-initialization upon request
  - Ability to upload its configuration
  - Ability to command half-routers to establish and terminate connections.

5.5.2.3 BACnet Advanced Application Controller

A BACnet Advanced Application Controller (B-AAC) is a control device with limited resources compared to a B-BC. It is intended for specific applications that do not require that much programming. For information on a B-AAC's functionality, refer to the BACnet standard. Figure 5.65 shows an example of a B-AAC.

Fig. 5.65 A BACnet advanced application controller [5]

5.5.2.4 BACnet Application Specific Controller

The BACnet Application Specific Controller (B-ASC) is a controller designed for specific applications and has limited resources compared to B-AAC. A B-ASC is normally programmed by the manufacturer, which means the user can only change the parameters.

5.5.2.5 BACnet Smart Actuators and BACnet Smart Sensors

BACnet smart actuators (B-SA) and smart sensors (B-SS) are simple devices that can be communicated with using BACnet. Their resources are limited to sending values as electrical signals upon request.

5.5.2.6 BACnet Routers

BACnet routers are devices that interconnect networks that have the same or different network technology. The controller will often have integrated "router" functionality. This means that the B-AAC shown in Fig. 5.65 can switch between MS/TP and Ethernet.

5.5.3 Protocol Implementation Conformance, Conformance Test and Certification of BACnet Devices

BACnet devices must conform precisely to the requirements of the BACnet standard. To show that a device conforms to the BACnet protocol, the manufacturer must supply a Protocol Implementation Conformance Statement (PICS) that identifies all the sections of BACnet that are implemented in the device.
A BACnet PICS should contain the following information:

- Vendor, product ID, software version
- The standardized BACnet device profile to which the device conforms (if any)
- The BIBBS supported by the device
- Network capabilities
- Object types supported by the device
- Any optional properties supported
- Other features

The BACnet devices must then pass a conformance test verifying the correct implementation of the standard object types and services indicated in the PICS. The PICS enables you to determine at the planning stage whether certain devices are interoperable.

The conformance test covers all the functions and OSI layers, from accessing individual objects through to the data link layer protocols that are described in the PICS. The BACnet Testing Laboratory (BTL) is an independent institution established by BACnet International to support compliance testing and interoperability testing activities. Once a product has passed all the tests it is awarded the BTL Mark.

5.6 Gateways to Other Systems

As well as the use of routers to interconnect two or more BACnet networks, gateways also play a key role because they allow you to connect systems that use different technologies (Fig. 5.66) such as BACnet and KNX or BACnet and LonWorks.

A BACnet gateway converts messages in one protocol into messages in another protocol. The term "gateway," however, can have a different meaning. For example, the standard gateway in the Windows operating system's network parameters is nothing more than the IP address of the next router.

By way of comparison, BACnet gateways have to cope with a difficult task because the concepts and data structures of the different automation systems are often not compatible. In this respect, BACnet is very flexible and its many objects and services enable it to be customized to "foreign" systems.

BACnet is often used as the superordinate system. When choosing a gateway the question arises, you need to determine what information and functionalities of the foreign system are available in BACnet. You also need to establish whether you just want to request data from or control devices in the other system. This means you must determine the capabilities of a gateway in detail.

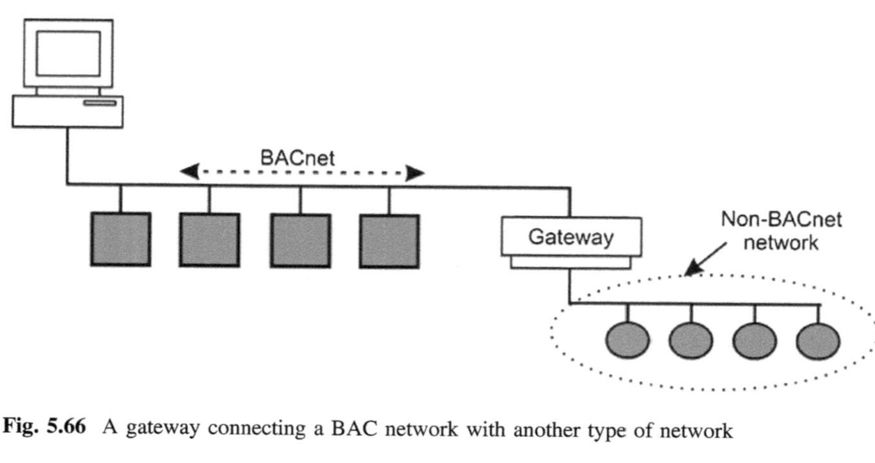

Fig. 5.66 A gateway connecting a BAC network with another type of network

5.7 Exercises

Exercise 5.1
What does the abbreviation BACnet stand for?

Exercise 5.2
What advantages do open protocols have over their proprietary counterparts?

Exercise 5.3
At which levels in building automation are BACnet, LonWorks and KNX used?

Exercise 5.4
Why does BACnet only have four layers instead of the seven in the OSI model?

Exercise 5.5
What is data encapsulation?

Exercise 5.6
What is the effect of the transfer rate and the response time on the choice of technology selected for the field and management levels?

Exercise 5.7
Why is the EIA-485 standard used for MS/TP and not EIA-232?

Exercise 5.8
Which topology is used in EIA-485 networks?

Exercise 5.9
A network consists of five computers (A to E) and other network components (hub, switch). The following table shows which computers see the frames that are transported between the computers listed in the column on the far left-hand side. Assume that the network has been up and running for a while so that the switch knows which computers are connected to it.

Data transmission from	A	B	C	D	E
A to E	x		x	x	x
B to E		x			x
E to A	x		x	x	x

Sketch the possible networks. Mark the collision domains and the broadcast domains.

Exercise 5.10
Explain why local and remote collisions do not occur in a solely switch-based network.

Exercise 5.11
Collision domains are either smaller or the same size as broadcast domains. True or false?

Exercise 5.12
Compare the different types of twisted-pair cable with regard to electromagnetic compatibility (influence or creation of interference) and ease of installation.

Exercise 5.13
What is the advantages and disadvantages of multi-level transmission technology?

Exercise 5.14
What is the difference between single-mode and multimode fiber with regard to dispersion?

Exercise 5.15
Compare the transmission rate and maximum distance of twisted-pair and fiber-optic cables.

Exercise 5.16
With the help of the Internet, find out the name of the manufacturer of the network card with the MAC address 0 × 00A02433F217.

Exercise 5.17
You can record the sending and receiving of frames using protocol analyzers such as Wireshark [www.wireshark.org]. After you have consulted your network administrator, install the Wireshark and examine the structure of the sent frames based on Fig. 5.31. The preamble is not shown.

Exercise 5.18
What are the advantages and disadvantages of hierarchical and flat addresses?

Exercise 5.19
What are the differences between public and private IP addresses?

Exercise 5.20
Complete the following table.

A computer's IP addresses	Network class	Subnet mask	Network address	Broadcast address	Max. number of computers in the subnet
100.100.100.20		255.255.255.0			
200.200.1.253			200.200.1.252	200.200.1.255	
143.93.64.7					126
17.3.172.43					1022

Exercise 5.21
Using the Internet, find out which ports and transport protocols are used for sending and receiving e-mails.

Exercise 5.22
Data packets N bytes in length are to be transported over a bidirectional pathway (transmission rate = 2 Mbit/s, simple signal transit time of 100 ms).

No transmission errors occur. The receiver acknowledges the correct packets immediately after they arrive using only one 1 bit (i.e., the duration is negligible). Using the stop-and-wait method, what is the throughput to transmission rate ratio (as a percentage) for the following packet lengths: $N = 500$ bytes and $N = 1500$ bytes? What other method would achieve a higher throughput?

Exercise 5.23
Explain the difference between TCP and UDP. Why does UDP not need a three-way handshake?

Exercise 5.24
What are the advantages and disadvantages of tunneling routers and BACnet/IP?

Exercise 5.25
Using the numbering system in Tables 5.9 and 5.10, what is the BACnet network number and the object identifier for device 18 on the 10th floor in building 123.

Exercise 5.26
Which object can be used to change the properties of several other objects at the same time?

Exercise 5.27
Which objects would you use to write scheduling information to a BACnet device's memory?

Exercise 5.28
The average temperature in a cooler over a period of 2 h is to be calculated and recorded. You have a temperature sensor with a voltage output channel. The recorded data is to be kept for 1 week, after which it is deleted. What objects will you need? List their typical properties.

Exercise 5.29
What does "deadband" refer to in conjunction with threshold notifications?

Exercise 5.30
Which object can be used to control proprietary application programs?

Exercise 5.31
Expand exercise 5.28 so that an alarm is raised when the setpoint temperature of –20 °C is exceeded by 5 °C. This alarm should set off siren 1 (binary controllable) on work days and siren 2 (also binary controllable) at weekends. Which objects with what properties will you need?

Exercise 5.32
Which notification levels and priority ranges would be assigned to the following events: detection of a broken window, blind is stuck, the pump on the waste water pump station has broken down, and the filter in the ventilation system is blocked?

Exercise 5.33
Complete Table 5.44 for the following scenario: During the minimum OFF time, an ON command is received with a priority level of 4. What happens when a command with priority 7 arrives before the minimum OFF time has elapsed?

Exercise 5.34
What is the difference between an IA and a BIBB?

Exercise 5.35
Explain why B-OWS does not contain a SCHED-I-B BIBB.

Exercise 5.36
Arrange B-ASC, B-SS, B-BC, B-SA and B-AAC according to increasing functionality.

Exercise 5.37
Why should you avoid using a gateway if possible? Give reasons.

References

1. VDI-TGA/BIG-EU-Leitfaden zur Ausschreibung interoperabler Gebäudeautomation auf der Basis von DIN EN ISO 16484-5 Systeme der Gebäudeautomation—Datenkommunikationsprotokoll BACnet, 2009 www.big-eu.org
2. Kranz HR (2013) BACnet Gebäudeautomation 1.12. cci Dialog, Karlsruhe
3. Tanenbaum AS, Weatherall DJ (2012) Computernetzwerke. Pearson Studium, Hallbergmoos
4. DIN EN ISO 16484-5 (2014) Systeme der Gebäudeautomation—Teil 5: Datenkommunikationsprotokoll (ISO 16484-5: 2014); Englische Fassung EN ISO 16484-5: 2014. Beuth, Berlin
5. Andover-Continuum-BACnet-b3920-Datasheet.pdf (2011) www.schneider-electric.com

Index

H. Merz et al., *Building Automation*, Signals and Communication Technology,
https://doi.org/10.1007/978-3-319-73223-7

S

T

U

MIX
Papier aus verantwortungsvollen Quellen
Paper from responsible sources
FSC® C105338

If you have any concerns about our products,
you can contact us on
ProductSafety@springernature.com

In case Publisher is established outside the EU,
the EU authorized representative is:
Springer Nature Customer Service Center GmbH
Europaplatz 3, 69115 Heidelberg, Germany

Printed by Libri Plureos GmbH
in Hamburg, Germany